Knots and Links

Knots and links are studied by mathematicians, and are also finding increasing application in chemistry and biology. Many naturally occurring questions are often simple to state, yet finding the answers may require ideas from the forefront of research.

This readable and richly illustrated book explores selected topics in depth in a way that makes contemporary mathematics accessible to an undergraduate audience. It can be used for upper-division courses, and assumes only knowledge of basic algebra and elementary topology. The techniques developed include combinatorics applied to diagrams, the study of surface intersections and 'cut and paste' surgery, skein theory of polynomial invariants, and the properties of tangles. Together with standard topics such as Seifert matrices and Alexander and Jones polynomials, the book explains polygonal and smooth presentations, the surgery equivalence of surfaces, the behaviour of invariants under factorisation and the satellite construction, the arithmetic of Conway's rational tangles, and arc presentations. The families of torus knots, pretzel knots, rational (or 2-bridge) links and doubles of the trefoil recur as examples throughout the text.

Alongside the systematic development of the main theory, there are discussion sections that cover historical aspects, motivation, possible extensions and applications. Many examples and exercises are included to show both the power and the limitations of the techniques developed.

The construction of a braided bidirectional satellite. The vertical line is an axis that provides the reference frame for the structure. The transparent tube is a torus knotted in the form of a trefoil and is the companion torus of the satellite. Its arrangement is based on an arc presentation of the companion knot: an assembly of five semicircles meeting at the axis (see Figure 10.6). The two loops inside the torus form the satellite link. They run around the torus in opposite directions (the satellite is bidirectional), but they run around the axis in the same direction (so they are braided). (Image created with PovRay.)

Knots and Links

Peter R. Cromwell

CAMBRIDGE
UNIVERSITY PRESS

PUBLISHED BY THE PRESS SYNDICATE OF THE UNIVERSITY OF CAMBRIDGE
The Pitt Building, Trumpington Street, Cambridge, United Kingdom

CAMBRIDGE UNIVERSITY PRESS
The Edinburgh Building, Cambridge CB2 2RU, UK
40 West 20th Street, New York, NY 10011-4211, USA
477 Williamstown Road, Port Melbourne, VIC 3207, Australia
Ruiz de Alarcón 13, 28014 Madrid, Spain
Dock House, The Waterfront, Cape Town 8001, South Africa

http://www.cambridge.org

First published 2004

Printed in the United Kingdom at the University Press, Cambridge

This book was typeset by the author using Latex

A catalogue record for this book is available from the British Library

Library of Congress Cataloguing in Publication data

Cromwell, Peter R., 1964–
   Knots and links / Peter R. Cromwell.
       p.   cm.
   Includes bibliographical references and index.
   ISBN 0 521 83947 5 (hardback) – ISBN 0 521 54831 4 (paperback)
   1. Knot theory.   2. Link theory.   I. Title

QA612.2.C75   2004
$514'.2242$ – dc22   2004045914

ISBN 0 521 83947 5 hardback
ISBN 0 521 54831 4 paperback

Contents

Preface

Knot theory is the study of embeddings of circles in space. It is a subject in which naturally occurring questions are often so simple to state that they can be explained to a child, yet finding answers may require ideas from the forefront of research. It is a subject of both depth and subtlety.

The subject started to develop systematically in the late nineteenth century, and has seen explosive growth during the last 20 years. Even so, many of the ideas and recent results can be explained without the use of advanced mathematical technology and can be made accessible to undergraduates.

The study of knots and links is part of topology, the branch of geometry that deals with flexible and deformable spaces. A topologist concentrates on properties that remain unchanged by continuous transformations that can be undone. In a first topology course students meet:

- topological spaces and their main properties — open and closed sets, continuity, homeomorphisms, connectedness, compactness;
- topological structures — concrete topology, discrete topology, metric spaces;
- topological constructions — product spaces, quotient spaces;
- topological invariants like the Euler characteristic;
- examples such as the classification of surfaces.

These are all intrinsic properties of a space. Another aspect of topology is the study of the ways that spaces can sit inside one another — the embedding problem. Knot theory is a perfect introduction to this and makes a good second course in topology.

This book gives an introductory account of knot theory from the geometric viewpoint. It provides a good grounding in the subject for anyone wishing to do research in knot theory or geometric topology, but also stands on its own as a course suitable for final year undergraduates.

Philosophy

Many courses for undergraduate students teach them a particular toolkit. Students can become proficient at using the tools, but they see very few applications. This often continues in their final year. Most students do not continue on to postgraduate study so it seems inappropriate to give them yet more specialised tools they will never need. This book is not tool-driven. Rather, it focusses on the properties of the objects of study (knots, links, and surfaces). It takes the view that students can explore a topic in some depth by applying the tools they already have and developing new ones along the way. It is the subject under investigation that motivates the development of new tools, not the tool-set that constrains the choice of examples.

Content

Classical knot theory is a part of 3-dimensional geometric topology. However, in recent years the scope of the subject has expanded enormously and it now has connections with singularity theory (the finite type invariants of Vassiliev and Goussarov), symplectic geometry (Legendrian knots from contact structures) and theoretical physics (quantum groups and statistical mechanics). With such a wealth of material, it is tempting to try to include a little of everything, making frequent digressions to introduce definitions and quote theorems. I have tried to avoid this path and to build a coherent picture by concentrating on a few topics in detail. My bias is toward geometry and combinatorics. The techniques developed include

- combinatorics applied to diagrams,
- the study of surface intersections and 'cut and paste' surgery,
- skein theory of polynomial invariants,
- properties of tangles.

Some homology theory is introduced when it is needed, but (in line with the underlying philosophy) it is only developed so far as is necessary to do the job required. It is used to describe the basis for H_1 of a surface, leading to the linking form, the Seifert matrix and invariants derived from it; it also features in the discussion of linking numbers, and the classification of surface automorphisms.

Even though attention has been restricted to the classical arena, the book includes recent developments. Contributions are continually being

made, both new results and improvements in technique. I have tried to
collect here elegant, concise or informative proofs. This makes some difficult
results accessible for the first time (for example, the surgery equivalence of
surfaces). The book also includes a detailed discussion of Conway's rational
tangles, a topic that is often mentioned in passing but whose explanation
tends to be glossed over.

Alongside the development of the main theory, there are some discussion
sections that cover historical aspects, motivation, possible extensions, and
applications.

I should also mention topics that are omitted. Apart from basic homology theory, concepts from algebraic topology such as the fundamental group
are absent: their application to knot theory is well documented in other
sources [9, 15]. Also excluded are discussions of colouring diagrams (see [13]
for a good treatment), periodicity, surgery constructions of 3-manifolds,
covering spaces, fibrations, and the more recent finite type invariants and
quantum invariants.

Approach

It is hard to make an introduction to knot theory truly self-contained. In
some accounts, the problems of wild embeddings and methods for dealing
with them are concealed from the student. Here the problems are discussed
and several equivalent ways to define tameness are presented: polygons,
smooth or locally flat embeddings, and diagrams with finitely many crossings. To develop the subject rigorously requires the foundations provided
by the development of 3-dimensional geometric topology: we need to quote
theorems of piecewise linear topology, general position, and so on. These
tools are collected together in Chapter 2. We regard them as axioms for
the development of the subject. Postgraduate students who have seen the
great benefits that can be gained by their application may want to study
the underlying theory.

Apart from these foundations, I have adopted the view that statements
should be proved; the few external references that are included are of a
supplementary nature and not essential to the development. The subject is
developed in a systematic and rigorous fashion. Knot theory has a strong
visual element and many diagrams are used in the explanations.

An underlying goal used to monitor progress throughout the book is to
distinguish all the knots in Appendix A. The book includes many examples,
some in the form of exercises, which show both the power and the limitations

of the techniques developed. The families of torus knots, pretzel knots, rational (or 2-bridge) links, and doubles of the trefoil recur as examples throughout the text. The behaviour of invariants under factorisation and the satellite construction is also strongly featured.

Prerequisites

I have assumed that the student is familiar with the following tools, which are covered in the first years of a normal undergraduate mathematics course:

- topology (as listed above);
- graph theory;
- linear algebra — matrices, determinant, signature, change of basis;
- algebra — groups, modules, and their quotients;
- mathematical literacy — set notation, equivalence relations and equivalence classes, basic terminology (associative, commutative, reflexive, transitive, *etc.*).

No background in algebraic topology is required.

Acknowledgements

I am very grateful to many people who have provided information, encouragement, and feedback on early versions of the book, including: Jim Hoste, Vaughan Jones, Louis Kauffman, Pedro Manchon, Hugh Morton, Hitoshi Murakami, Jozef Przytycki, Brian Sanderson, and the members of the KOOK seminar in Japan. I must make special mention of Elisabetta Beltrami who provided many detailed and valuable comments. Her careful thought and attention have made a significant improvement to the clarity and accuracy of the presentation.

Peter Cromwell   Liverpool, 2004

Notation

Set notation

$\varnothing$	empty set
$\cap$	intersection
$\cup$	union
$\subset$	is a subset of
$\supset$	contains (as a subset)
$\in$	is a member of
$\ni$	contains (as a member)
$|S|$	number of elements of the set S
$X - Y$	$\{x \in X : x \notin Y\}$, complement of Y in X

Standard spaces

$\mathbb{R}^n$	n-dimensional Euclidean space
S^n	n-dimensional sphere
$\mathbb{H}^n$	n-dimensional hyperbolic space

Algebraic structures

$\mathbb{N}$	set of natural numbers, starting at 1
$\mathbb{N}_0$	set of natural numbers, starting at 0
$\mathbb{Z}$	ring of integers
$\mathbb{Q}$	field of rational numbers
$\mathbb{R}$	field of real numbers
$\mathbb{C}$	field of complex numbers
$\mathbb{B}_n$	group of n-string braids
$\mathbb{Z}[t]$	ring of polynomials in t with integer coefficients
$\mathcal{L}(M)$	linear skein of 3-manifold M under some skein relation

Links

For knots up to ten crossings and links up to nine crossings I use the classical notation of Rolfsen's catalogue ([5], Appendix C). For higher order knots I give the entry in the *Knotscape* database [254] in the form of two numbers separated by a letter: the first number is the number of crossings, the second number is the index, the letter is 'a' for alternating knots and 'n' otherwise. Unfortunately, where the catalogues overlap, the indices do not agree.

Note that 10_{161} and 10^*_{162} in [5] are actually the same knot. Because of this, some authors now renumber the last four knots 10_{163}–10_{166} in the catalogue, reducing the index by one. I have not done this.

K	knot
L	link
L^*	mirror image of L
$-L$	link L with orientation of all components reversed
$\|L\|$	link with orientation ignored
$L_1 \sqcup L_2$	distant union of L_1 and L_2 (split link)
$L_1 \# L_2$	link with factors L_1 and L_2
$\bigcirc$	trivial knot
$T(p,q)$	(p,q) torus knot
$P(p,q,r)$	(p,q,r) pretzel link

Link invariants

$\alpha(L)$	arc index of L
$b(L)$	bridge index of L
$c(L)$	crossing number of L
$\det(L)$	determinant of L
$g(L)$	genus of L
$\mu(L)$	multiplicity (number of components) of L
$n(L)$	nullity of L
$p(L)$	polygon index of L
$s(L)$	braid index (or Seifert circle index) of L
$\sigma(L)$	signature of L
$u(L)$	unknotting number of L
$\Delta_L(x)$	Alexander polynomial of L
$\nabla_L(z)$	Conway polynomial of L
$V_L(t)$	Jones polynomial of L
$F_L(a,x)$	Kauffman polynomial of L
$P_L(v,z)$	Homfly polynomial of L

Others

M^{T}	transpose of matrix M
D	link diagram
$c(D)$	crossing number of diagram D
$w(D)$	writhe of diagram D
$\langle D \rangle$	Kauffman's bracket polynomial of diagram D
λ	loop
Δ	disc
F	surface
∂F	boundary of surface F
$\mathrm{int}(F)$	interior of surface F
$g(F)$	genus of surface F
$\chi(F)$	Euler characteristic of surface F
$|X|$	number of connected components of space X
$\mathrm{maxdeg}\, p()$	maximum degree of one-variable polynomial p taken over terms with non-zero coefficients
$\mathrm{mindeg}\, p()$	minimum degree of polynomial p
$\mathrm{breadth}\, p()$	$\mathrm{maxdeg}\, p() - \mathrm{mindeg}\, p()$ for polynomial p
$x\text{-}\mathrm{maxdeg}\, p()$	maximum degree in variable x of multi-variable polynomial p taken over terms with non-zero coefficients
$x\text{-}\mathrm{mindeg}\, p()$	minimum degree in variable x of polynomial p
$x\text{-}\mathrm{breadth}\, p()$	$x\text{-}\mathrm{maxdeg}\, p() - x\text{-}\mathrm{mindeg}\, p()$ for polynomial p

References

In the text or figure captions a number between square brackets refers to an entry in the bibliography.

1 Introduction

We are all familiar with everyday knots — the kind that we use to tie up parcels, shoelaces, and so on. These knots can be untied, and retied in the same or different ways. By manipulating the string we can let the 'knots' escape.

To make a study of knottedness, the knotted part of the string must be trapped. One way to do this is to imagine an infinitely long string which is a straight line outside the region containing the knot (see Figure 1.1). A simpler way is to join the ends to form a loop.

Knot theory is the study of the topological properties of loops embedded in 3-dimensional space. Mathematical knots are modelled on the physical variety, and we allow a knot to be deformed as if it were made of a thin, flexible, elastic thread.

We shall begin by making these ideas more precise.

1.1 Knots

We shall see later that the following definition is a bit too general and that something more subtle is required. However, it is a good place to start.

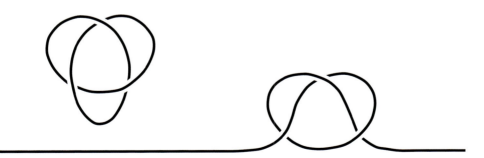

Figure 1.1. A knot can be trapped in a loop or an infinitely long string.

Definition 1.1.1 (knot).   A *knot* $K \subset \mathbb{R}^3$ is a subset of points homeomorphic to a circle.

Some examples are shown in Figure 1.2. Clearly, a first example is a circle — a planar, round circle. We take this closed curve to be the archetypal unknotted loop and call it the *trivial* knot. The overhand knot commonly used to tie string is called a *trefoil*. It comes in two mirror-image forms labelled left-handed and right-handed. Tying two overhand knots in succession produces either a granny knot or a reef knot, depending on which kinds of trefoils are used. The knots shown here are so common that they

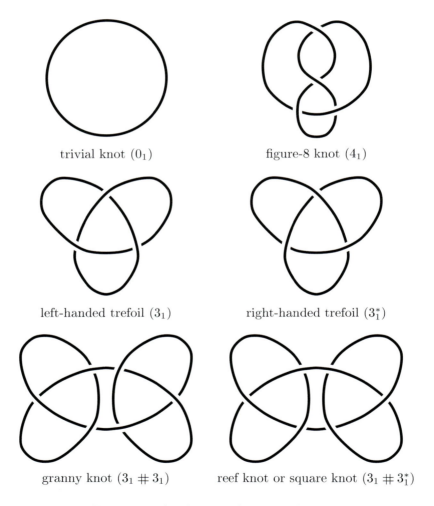

trivial knot (0_1)   figure-8 knot (4_1)

left-handed trefoil (3_1)   right-handed trefoil (3_1^*)

granny knot ($3_1 \,\#\, 3_1$)   reef knot or square knot ($3_1 \,\#\, 3_1^*$)

Figure 1.2. A selection of common knots.

have names. Many more knots are identified only by a catalogue number. Two catalogues are used in this book. Labels of the form N_m refer to the classical hand-made catalogue of small knots, part of which is reproduced in Appendix A. The catalogue is organised in order of increasing complexity: N is called the *order* of the knot and N_m is just the mth knot of order N in the list. Labels of the form $N*m$ where the '*' is either 'a' or 'n' refer to the much larger computer-generated *Knotscape* catalogue. (See page xvi for further comments.) The trivial knot has label 0_1 but we shall also denote it by $\bigcirc$.

Notice that the definition of a knot says nothing about how the circle should be arranged in space. Figure 1.3 shows two possibilities: a polygon and a smooth curve. We shall explore examples of both of these later.

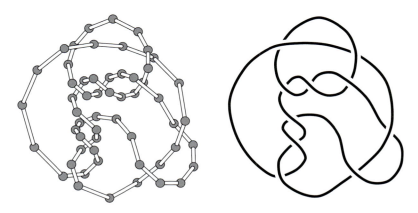

Figure 1.3. A polygonal and a smooth representation of the knot 10_{50}.

1.2 Deformation

Like knots in string, we want mathematical knots to be flexible. We need to model the physical manipulation by which a thread can be deformed into different positions.

If you have learned any algebraic topology, you will be familiar with the deformation known as homotopy. A *homotopy* of a space $X \subset \mathbb{R}^3$ is a continuous map $h : X \times [0, 1] \longrightarrow \mathbb{R}^3$. The restriction of h to level $t \in [0, 1]$ is $h_t : X \times \{t\} \longrightarrow \mathbb{R}^3$. Here, t is chosen deliberately to indicate time, and the images $h_t(X)$ for increasing values of t show the evolution of X in $\mathbb{R}^3$. We require that h_0 is the identity map. All the h_t's inherit the continuity property of h.

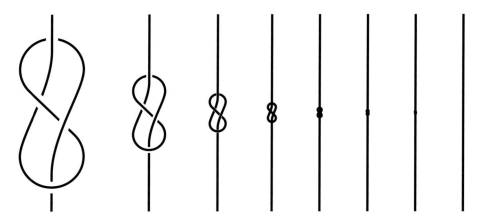

Figure 1.4. Bachelors' unknotting — a continuous transformation that makes knots vanish.

Homotopy is no use for deforming knots because it allows a curve to pass through itself: all knots are homotopic to the trivial knot. We need to ensure that each h_t is one-to-one. If h_t has this property for all $t \in [0,1]$ then h is called an *isotopy*.

Unfortunately, isotopy is no use for deforming knots either. Even though a knot cannot pass through itself under isotopy, all knots are isotopic to the trivial knot. To see why this is so, imagine pulling on the thread really hard so that the knot becomes very small and tight. Because mathematical thread has no thickness, we can reduce a mathematical knot until it becomes a point and disappears. This continuous transformation, known as bachelors' unknotting, is sketched in Figure 1.4.

We need only make one minor modification to fix this problem: we just need to ensure that the space containing the knot moves continuously along with the knot and does not become irreparably contorted. One way to visualise what is required is to imagine the knot embedded in a viscous syrup; stirring the syrup causes the knot to move in response.

An ambient isotopy of a space $X \subset \mathbb{R}^3$ is an isotopy of $\mathbb{R}^3$ that carries X with it. This is what we need to model the topological deformation of knots.

Definition 1.2.1 (isotopy equivalence). Two knots K_1 and K_2 are *ambient isotopic* if there is an isotopy $h : \mathbb{R}^3 \times [0,1] \longrightarrow \mathbb{R}^3$ such that $h(K_1, 0) = h_0(K_1) = K_1$ and $h(K_1, 1) = h_1(K_1) = K_2$.

Examples of knots which are ambient isotopic are shown in Figure 1.5.

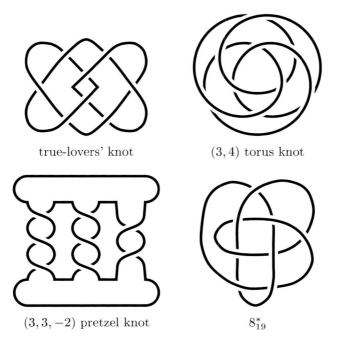

true-lovers' knot $(3, 4)$ torus knot

$(3, 3, -2)$ pretzel knot 8_{19}^{*}

Figure 1.5. These knots are all ambient isotopic.

Ambient isotopy is an equivalence relation on knots: two knots are equivalent if they can be deformed into one another. Each equivalence class of knots is called a *knot type*; equivalent knots have the same type.

However, this rather formal distinction between a knot and its type is often forgotten. With some abuse of terminology, the word 'knot' is applied to mean a whole equivalence class (a knot type) or a particular representative member whose properties we are interested in. In general, this does not cause any misunderstanding. For example, when we say two knots are different, we actually mean that they are inequivalent (have different types). If a loop has the same type as the trivial knot, we say it is *unknotted*.

1.3 Polygonal knots

So far we have only looked at knots in general terms. How can we specify a particular knot? One method is to think of it as built up of straight lines and give the coordinates of the corners.

A spatial *polygon* is a finite set of straight line segments in $\mathbb{R}^3$ which intersect only at their endpoints; the lines are called *edges*, and their endpoints

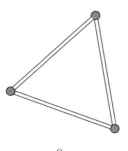

0_1

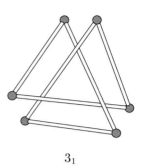

3_1

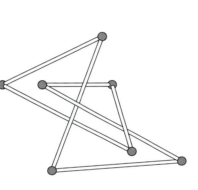

4_1

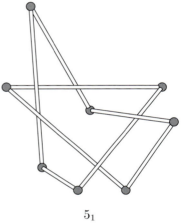

5_1

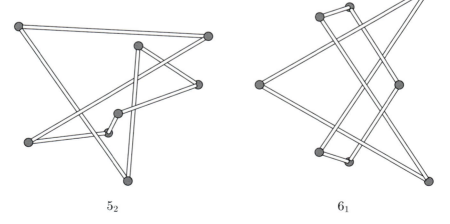

5_2

6_1

Figure 1.6. The twelve knots that can be made with at most eight edges.

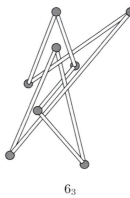

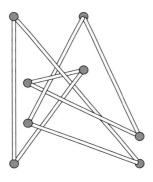

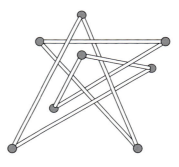

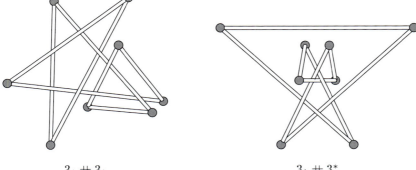

6_2

6_3

8_{19}

8_{20}

$3_1 \# 3_1$

$3_1 \# 3_1^*$

Figure 1.6 (*continued*).

are called *vertices*; exactly two edges meet at every vertex. The picture on the left of Figure 1.3 shows a 'ball and stick' model of a knotted polygon. More examples of polygonal knots are given in Figure 1.6: the ordered lists of vertices defining each one are given in Appendix E. The knots are, in fact, all different, but we shall not prove this until much later.

There are two basic questions we can ask about polygonal knots:

1. How many edges do I need to make a given knot K?

2. What knots can I make using n edges?

The first thing to say is that question 1 presupposes that a given knot *can* be represented as a polygon. At the end of the chapter we shall explore whether this is true. Many of the pictures of knots in this book appear to have been drawn with smooth lines, but we can interpret them as polygons which have lots of very short edges. (In reality, this is how the computer draws them.) Once we have found a polygonal representation of the knot we are interested in, we can manipulate it and try to reduce the number of edges. The polygon in Figure 1.3 has over 50 edges and this is clearly more than necessary. If we repeatedly perturb the vertices (making sure the knot type is preserved), some edge may shrink until it is so short that it can be collapsed and removed without affecting the knot type. Computer programs that let knots evolve in this way have been used to find candidates for minimal polygons.

The answer to question 2 is shown in Figure 1.6 for $n = 3$, 6, 7 and 8. There must be at least three edges to make a triangle (the trivial knot), and it is not hard to show that a knotted polygon must have at least six edges (Exercise 1.12.7). Enumerating these polygons soon becomes impracticable because the number of cases explodes as n increases.

1.4 Smooth knots

Recall that a function which can be differentiated r times and whose rth derivative is continuous is called a C^r function. A C^∞ function can be differentiated as many times as we like. A function which is C^r for some $r > 0$ is called *smooth*.

Smooth functions provide another method to specify particular knots. For example, the function

$$
\begin{aligned}
f : \mathbb{R} &\longrightarrow \mathbb{R}^3 \\
t &\longmapsto (\sin(t),\, \cos(t),\, 0)
\end{aligned}
$$

maps the real line onto the unit circle in the xy-plane. The function is periodic and covers the circle infinitely many times. Note that it is the *image* of the function which is of interest to us, not the function itself.

In general we want a function

$$
\begin{aligned}
f : \mathbb{R} &\longrightarrow \mathbb{R}^3 \\
t &\longmapsto (x(t),\, y(t),\, z(t))
\end{aligned}
$$

in which the coordinate functions x, y and z are smooth functions of t, and f is periodic with minimum period τ so that $f(t+\tau) = f(t)$ for all t.

1.5 Torus knots

Some of the simplest knots to describe parametrically are torus knots, so-called because they lie in the surface of a standard torus.

We can generate a torus by taking a circle in the yz-plane of radius r, centred on the y-axis at distance $R + r$ from the origin, then rotating it about the z-axis. If we parametrise the circle by angle $\theta \in [0, 2\pi]$, and the rotation by angle $\phi \in [0, 2\pi]$, we can express the torus as

$$
\overset{\text{rotation}}{\begin{pmatrix} \cos(\phi) & -\sin(\phi) & 0 \\ \sin(\phi) & \cos(\phi) & 0 \\ 0 & 0 & 1 \end{pmatrix}} \overset{\text{circle}}{\begin{pmatrix} 0 \\ R + r\cos(\theta) \\ r\sin(\theta) \end{pmatrix}} = \overset{\text{torus}}{\begin{pmatrix} -\sin(\phi)\,(R + r\cos(\theta)) \\ \cos(\phi)\,(R + r\cos(\theta)) \\ r\sin(\theta) \end{pmatrix}}.
$$

The parameters r and R control the geometry of the torus: r is the radius of the tube and R is the radius of the hole. The angles form a coordinate system: any point on the torus can be labelled by a pair of the form (θ, ϕ).

The subset of points defined by the equation $p\,\phi = q\,\theta$ for coprime integers p and q winds its way around the torus and forms a knot. We need to solve the equation modulo 2π, or alternatively use the function

$$
t \longmapsto (-\sin(q\,t)\,(R + r\cos(p\,t)),\ \cos(q\,t)\,(R + r\cos(p\,t)),\ r\sin(p\,t)).
$$

This covers the curve more than once, but (as before) it is the image set that we are interested in, not the function. This knot is called the (p, q) torus knot and is denoted by $T(p, q)$. An example is shown in Figure 1.7. Note that a torus knot need not be presented in this form: all the pictures in Figure 1.5 depict the (3,4) torus knot even though it is only recognisable as such in the top-right picture. Apart from some basic symmetries (see

Figure 1.7. The $(15, -4)$ torus knot.

Exercise 1.12.3), different values of p and q give distinct knots, as we shall show in Theorem 10.5.4.

There are two distinguished curves on the torus: the *meridian* is $T(1, 0)$ and the *longitude* is $T(0, 1)$.

There is a nice way to make polygonal embeddings of torus knots that uses the fact that a hyperboloid of one sheet is a ruled surface [38]. First we need a parametrisation of the hyperboloid. Choose an angle $\theta \in [0, \pi/2]$ and construct two points: $A = (\cos\theta, -\sin\theta, -1)$ and $B = (\cos\theta, \sin\theta, 1)$. The straight line through A and B is defined by $x = \cos\theta$, $y = z\sin\theta$. Rotating this line about the z-axis gives the hyperboloid $x^2 + y^2 - z^2\sin^2\theta = \cos^2\theta$. Let H_θ denote the annulus obtained by restricting z to the interval $[-1, 1]$. The boundary curves of the annulus are unit circles: $x^2 + y^2 = 1$; $z = \pm 1$. The union of two of these annuli with different values of θ is topologically a torus.

The (p, q) torus knot with $p > q \geqslant 2$ can be embedded in one of these tori as follows. Choose θ and ϕ such that

$$\frac{q}{p} \cdot \frac{\pi}{2} < \theta < \min\left\{\frac{\pi}{2}, \frac{q}{p}\pi\right\} \qquad \text{and} \qquad \phi = \frac{q}{p}\pi - \theta.$$

The knot will lie in the torus $H_\theta \cup H_\phi$. The vertices of the knot are

$$v_i = \begin{cases} (\cos(i\pi\frac{q}{p}), \sin(i\pi\frac{q}{p}), -1) & \text{if } i \text{ is even,} \\ (\cos((i-1)\pi\frac{q}{p} + 2\theta), \sin((i-1)\pi\frac{q}{p} + 2\theta), 1) & \text{if } i \text{ is odd,} \end{cases}$$

for $i \in \{0, \ldots, 2p\}$. Notice $v_0 = v_{2p}$. The p odd-numbered vertices are equally spaced around the top circle, the p even-numbered ones around the bottom. The $2p$ edges are $e_i = [v_{i-1}, v_i]$ for $i \in \{1, \ldots, 2p\}$. The p even-numbered edges lie in H_ϕ, the p odd-numbered edges lie in H_θ. An example is shown in Figure 1.8, where $(p, q) = (5, 3)$, $\theta = \frac{35}{100}\pi$ and $\phi = \frac{25}{100}\pi$.

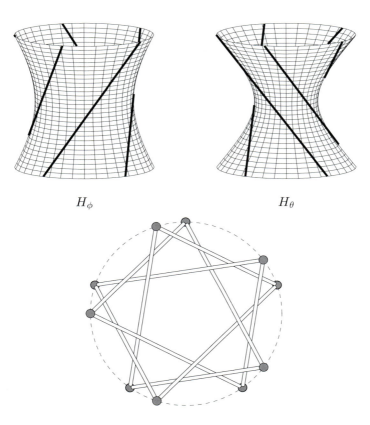

H_ϕ H_θ

Figure 1.8. The hyperboloid construction for polygonal torus knots [38].

1.6 Fourier knots and Lissajous knots

A Fourier series is an expression of the form

$$f(t) \; = \; \sum_{i=0}^{\infty} A_i \, \cos(B_i \, t + C_i)$$

where, for each term, $A_i \in \mathbb{R}$ is the amplitude, $B_i \in \mathbb{Q}$ is the frequency, and $C_i \in \mathbb{R}$ is the phase.

Suppose we are given a knot as a parametrised curve such that each of the coordinate functions is smooth (C^∞):

$$
\begin{aligned}
f : \mathbb{R} \; &\longrightarrow \; \mathbb{R}^3 \\
t \; &\longmapsto \; (x(t), \, y(t), \, z(t)).
\end{aligned}
$$

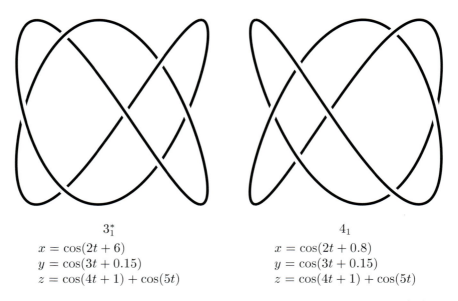

$$3_1^*$$

$$x = \cos(2t + 6)$$
$$y = \cos(3t + 0.15)$$
$$z = \cos(4t + 1) + \cos(5t)$$

$$4_1$$

$$x = \cos(2t + 0.8)$$
$$y = \cos(3t + 0.15)$$
$$z = \cos(4t + 1) + \cos(5t)$$

Figure 1.9. Fourier presentations of the trefoil and figure-8 knots [41].

Any periodic smooth function can be expressed as a Fourier series, and approximated arbitrarily closely by finitely many terms. Thus each of x, y and z can be expressed as a linear combination of cosine functions of various frequencies. Examples from [41] are shown in Figure 1.9; there are many more in [62].

The simplest case possible is when each coordinate function has just one term:

$$
\begin{aligned}
x(t) &= A_x \cos(B_x\, t + C_x), \\
y(t) &= A_y \cos(B_y\, t + C_y), \\
z(t) &= A_z \cos(B_z\, t + C_z).
\end{aligned}
$$

Knots which can be parametrised in this way are called *Lissajous* knots. In this case, the amplitudes are merely scaling factors in the coordinate directions so varying them does not change the topology of the knot. Similarly, it is only the differences between the phases which matter. Hence we can set $A_x = A_y = A_z = 1$ and $C_z = 0$. Furthermore, by reparametrising, we can assume that all the frequencies are coprime integers (Exercise 1.12.5).

Computer searches have been used to find knots with Lissajous presentations. Some examples from [24] are listed in the table below, and shown in Figure 1.10 viewed along the z-axis. More can be found in [45].

Knot	B_x	B_y	B_z	C_x	C_y
5_2	2	3	7	0.20	0.70
6_1	2	3	5	0.20	1.50
7_4^*	2	3	7	0.40	1.00
8_{15}	3	7	10	3.08	0.72
8_{21}^*	3	4	7	0.10	0.70
$3_1 \# 3_1^*$	3	5	7	0.70	1.00

By considering the relationship between $f(t)$ and $f(t + \pi)$, we can see that Lissajous knots have special (geometric) symmetry properties. If all the frequencies of a Lissajous knot are odd then reflection in the origin, $(x, y, z) \longmapsto (-x, -y, -z)$, carries the knot onto itself — it is a symmetry operation. If one of the frequencies, B_x say, is even then $(x, y, z) \longmapsto (x, -y, -z)$ is a symmetry. This is a 2-fold rotation about the x-axis and carries the knot onto itself (preserving its orientation); the knot is said to be periodic with period 2. There are some knots, 8_{10} for example, which are known not to possess these symmetries and hence cannot be Lissajous.

Other attributes associated with knots also take a special form on Lissajous knots. The Alexander polynomial (an invariant we shall meet in Chapter 7) can be used to show that a torus knot cannot have a Lissajous presentation [45]. However, there are infinitely many Lissajous knots (Exercise 9.8.9).

The cosine function used in the definition of Lissajous knots can be replaced by any function with period 2π that descends monotonically on $[0, \pi/2]$ and has the necessary symmetry properties: the knot type does not depend on the function used [40]. For example, we can replace the cosine function by the saw-tooth graph shown on the left of Figure 1.11. This produces a piecewise linear embedding which is isotopic to the standard Lissajous knot with the same coefficients. We can think of the polygon as the trajectory of a ball bouncing in a cube with perfect elastic collisions so that the angles of approach and return from the walls, floor and ceiling are equal. Polygons formed in this way are called *billiard* knots. An example is shown on the right of Figure 1.11.

1.7 Topological symmetries

Let $h : \mathbb{R}^3 \longrightarrow \mathbb{R}^3$ be a homeomorphism. A knot $K \subset \mathbb{R}^3$ and its image $h(K)$ are closely related. Indeed, if h is orientation-preserving then $h(K)$ is ambient isotopic to K.

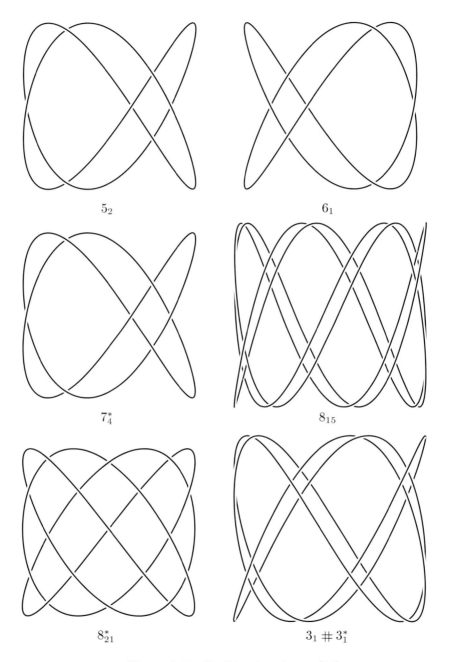

Figure 1.10. Six Lissajous knots [24].

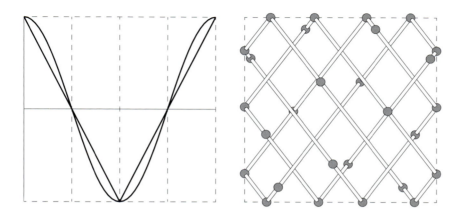

Figure 1.11. Replacing the cosine function by the saw-tooth function converts a Lissajous knot into a billiard knot. On the right is 8^*_{21}.

Suppose we use an orientation-reversing homeomorphism. Let $r : \mathbb{R}^3 \longrightarrow \mathbb{R}^3$ be the map $(x, y, z) \longmapsto (-x, -y, -z)$; this is called central inversion or reflection in the origin. The image $r(K)$ of a knot K is called the *mirror image, reflection,* or *obverse* of K and is denoted by K^*. (In other books you may see this written as $K!$ or $\overline{K}$.) If K and its mirror image are different then we say that K is *cheiral* (handed) and it occurs in a left-handed (*laevo*) and a right-handed (*dextro*) form. If $K = K^*$ then we say that K is *acheiral* or *amphicheiral*. The reflection of K is independent of the choice of r: the image of K under any other orientation-reversing homeomorphism will be ambient isotopic to $r(K)$.

Figure 1.12 shows a remarkable embedding of the figure-8 knot. This

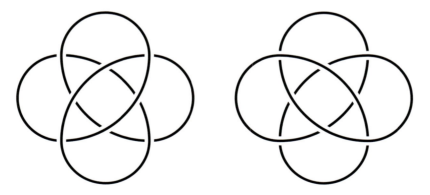

Figure 1.12. The figure-8 knot is amphicheiral — rotate the book 90 degrees.

knot is usually shown in a much simpler form, but in this configuration it is easy to see that it is amphicheiral. The diagram on the left can be obtained from that on the right by a reflection in the plane of the page — all the crossings are switched. It can also be obtained by rotating the diagram by 90 degrees. Thus the mirror image is not only isotopic to the original, the two are *congruent*.

So far we have ignored the possibility that a knot can be oriented. We orient a knot by nominating one of the two directions along it. If K is oriented then the knot with the opposite orientation, denoted by $-K$, is called the *reverse* of K. The knot $-K^*$ is called the *inverse* of K. Examples of four knot relatives are illustrated in Figure 1.13.

Let s denote the operation that sends K to $-K$, and let $t = rs$ denote the composite operation that sends K to $-K^*$. Since $-(-K) = K$, $(K^*)^* = K$, and $-(K^*) = (-K)^*$, the operations r, s and t form a group.

	1	r	s	t
1	1	r	s	t
r	r	1	t	s
s	s	t	1	r
t	t	s	r	1

The composition table above shows that any oriented knot which has two non-trivial symmetries must also have the third. Thus there are five possible topological symmetry types for oriented knots:

$$
\begin{aligned}
K = -K = K^* = -K^*, &\quad \text{fully symmetric;} \\
K = -K, &\quad \text{reversible;}^\star \\
K = K^*, &\quad +\text{amphicheiral;} \\
K = -K^*, &\quad -\text{amphicheiral;} \\
K, &\quad \text{asymmetric.}
\end{aligned}
$$

1.8 Links

So far we have considered embeddings of a single circle. It is natural to generalise this idea so that we can consider collections of such objects.

---

*This is often called invertible. Originally, $-K$ was called the inverse of K but this was changed when it was discovered that K and $-K^*$ are inverse elements in the knot cobordism group.

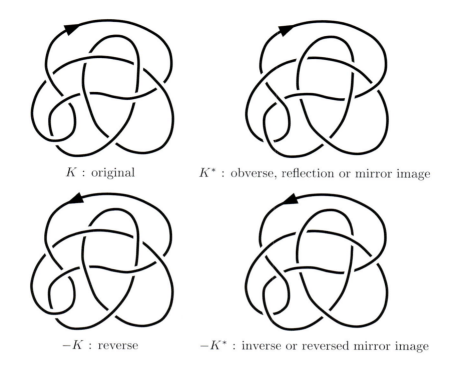

K : original K^* : obverse, reflection or mirror image

$-K$: reverse $-K^*$: inverse or reversed mirror image

Figure 1.13. The family of oriented knots derived from 9_{33}. They are all different but proving so is difficult.

Definition 1.8.1 (link). A *link* is a finite disjoint union of knots: $L = K_1 \cup \cdots \cup K_n$. Each knot K_i is called a *component* of the link. The number of components of a link L is called the *multiplicity* of the link, and is denoted by $\mu(L)$. A subset of the components embedded in the same way is called a *sublink*.

A set of n disjoint circles embedded in a plane is an example of the *trivial* link of multiplicity n. A link is trivial if all of its components are trivial knots and they are 'unlinked'. Equivalently, a link is trivial if its components bound disjoint discs.

Some simple examples of non-trivial links are shown in Figure 1.14. Each of the four 2-component links has two trivial knots as components, and so they can all be written as $0_1 \cup 0_1$. However, these four links are different so the notation $K_1 \cup \cdots \cup K_n$ is not enough to completely describe a link: it merely lists the component parts, and does not indicate how they are put together.

The classical catalogue numbering system is extended to links: a label of the form N_m^μ indicates the mth link with μ components of order N.

Clearly, the set of links contains the set of knots. We shall restrict the term knot to mean a link of only one component, but you should be aware that some authors use the terms knot and link interchangeably. Most of the time, we shall use the term 'links' to mean all links (including knots); in other cases, such as rational links or in the phrase 'knots and links', we shall be more restrictive and limit its scope to links with two or more components.

Links can be oriented: this means each component is assigned an orientation. An unoriented n-component link can be assigned orientations in 2^n ways.

Definition 1.8.2 (link equivalence). Two links L_1 and L_2 are *ambient isotopic* if there is an isotopy $h : \mathbb{R}^3 \times [0,1] \longrightarrow \mathbb{R}^3$ such that $h(L_1, 0) = h_0(L_1) = L_1$ and $h(L_1, 1) = h_1(L_1) = L_2$.

This definition of equivalence is *weak* in the sense that it does not impose any restrictions on the isotopy: there is a free choice of how to match up the components of L_1 with those of L_2. When we write a link as $K_1 \cup \cdots \cup K_n$, its components are given arbitrary labels $1, \ldots, n$. The components of a link may also be oriented. A stronger definition of equivalence requires the isotopy to preserve any labellings or orientations on the links.

Given any link L we can form its reflection L^*. Similarly, given an oriented link, we can form its reverse $-L$ by reversing the orientations on *all* of its components. In the same way as with knots, we say that a link is amphicheiral if $L = L^*$, and an oriented link is reversible if $L = -L$. Here we use '=' to mean equivalent in the weak sense.

When we studied the symmetries of oriented knots more closely, we discovered that there are five possible symmetry classes. With links, the situation becomes even more complicated. Suppose we have an oriented n-component link $L = K_1 \cup \cdots \cup K_n$. We denote the link formed from L by reversing the orientations of some components by $\pm K_1 \cup \cdots \cup \pm K_n$ where a $-$ sign indicates that the orientation of the component is reversed, and a $+$ sign that it is unchanged. We can also write L followed by the sequence of $+$ and $-$ signs, as is done in Figure 1.14. In this way we can form 2^n links corresponding to the 2^n n-tuples of signs $(\pm \ldots \pm)$. It is possible that some of these links will be strongly equivalent, so that corresponding components have the same labels and orientations. Furthermore, we can consider the

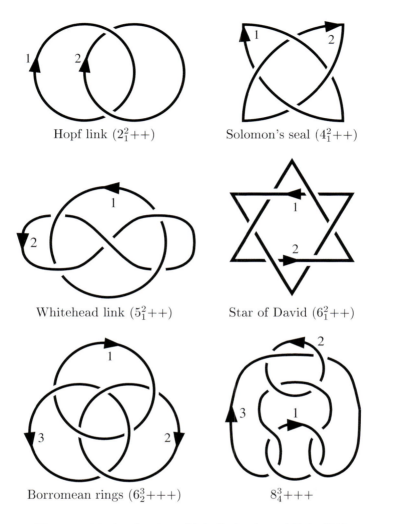

Figure 1.14. A selection of labelled oriented links [31].

strong equivalence of L^* to each of these 2^n links. If L^* is strongly equivalent to the link with sign vector ε, we can say that L is ε-amphicheiral. This is a natural extension of the notation $+$ amphicheiral and $-$ amphicheiral that we used for knots.

Let us look at some examples.

1. The Whitehead link.

The diagram in the centre-left of Figure 1.14 shows a 2-component link with orientations marked, and the components labelled 1 and 2. This is the labelled oriented link 5_1^2++. (The unlabelled unoriented link is denoted simply by 5_1^2.) We form the link 5_1^2-+ by reversing the orientation on the component labelled 1. Similarly for 5_1^2+- and 5_1^2--. Potentially, this gives us four distinct links. However, the four diagrams are all related by 2-fold rotations so there is only one isotopy type. This link is cheiral: $(5_1^2++)^* \neq 5_1^2++$.

2. The Hopf link.

The diagram in the top-left of Figure 1.14 shows the labelled oriented link 2_1^2++. As before, we can form diagrams 2_1^2-+, 2_1^2+- and 2_1^2-- by reversing orientations. In this case $2_1^2++ = 2_1^2--$ and $2_1^2+- = 2_1^2-+$. Furthermore, these two link types are mirror images: $(2_1^2++)^* = 2_1^2+-$. Therefore the Hopf link is $+-$ amphicheiral.

3. Solomon's seal.

As in the previous case, the four links divide into two isotopic pairs: $4_1^2++ = 4_1^2--$ and $4_1^2+- = 4_1^2-+$. Here, however, the two link types are not mirror images: $(4_1^2++)^*$ is different from both.

There are other 2-component links for which all four orientations are different; in some cases the four occur in two mirror-image pairs [28].

With more than two components there are even more possibilities. Cheirality can even depend upon the orientations chosen. For example 8_4^3+++ and 8_4^3++- are amphicheiral, but $(8_4^3+-+)^*$ is equivalent to 8_4^3+--.

1.9 Invariants

A major part of knot theory is the study of link invariants.

Definition 1.9.1 (link invariant). A *link invariant* is a function from the set of links to some other set whose value depends only on the equivalence class of the link. Any representative from the class can be chosen to calculate the invariant. There is no restriction on the kind of objects in the target space. For example, they could be integers, polynomials, matrices, or groups.

Probably the simplest link invariant is the multiplicity of a link: $\mu(L)$ is the number of components of L.

Although some invariants are used mainly to distinguish different links, many are related to geometric or topological properties and measure the complexity of a link in various ways. Some of the simplest to define, and often hardest to calculate, minimise a numerical property. One of the oldest link invariants is of this type: the *unknotting number* is the least number of times that a link must pass through itself to be transformed into a trivial link, and is denoted by $u(L)$. Although one of the most obvious measures of complexity, this invariant is still incredibly hard to calculate. A non-trivial knot K that can be unknotted with only one pass has $u(K) = 1$. For other knots, an upper bound can be found by experiment; the problem is to prove that you cannot achieve the same result with fewer passes. We shall see one method for doing this in Chapter 6.

Another elementary invariant, which has been studied only recently, is the *polygon index*, denoted by $p(L)$. This is the minimum number of edges (or equivalently vertices) needed to represent the link as a collection of polygons. It is clear that the polygon index of a trivial link is $3\mu(L)$ — we need a triangle for each component. The hyperboloid construction for torus knots (§1.5) shows that the (p,q) torus knot with $p > q \geqslant 2$ has polygon index at most $2p$; in fact, this has been proved to be minimal when $2q > p$ [38]. Minimal polygons for knot types with polygon index at most 8 are shown in Figure 1.6. These have been found by the exhaustion of possibilities [52].

We shall investigate these and other invariants throughout the book.

1.10   The creation of catalogues

The first catalogue of knots was produced in 1876 by Peter Tait [56]. It contained diagrams of the 15 knots up to order 7. To enumerate them, Tait had systematically drawn pictures of knots and then grouped them into classes of equivalent knots. He did not have the benefit of topological methods and so could not prove that his knots were distinct: "Though I have grouped together many widely different but equivalent forms, I cannot be absolutely certain that all those groups are essentially different one from another."

What motivated this pioneering investigation? Tait was interested in work of Hermann von Helmholtz on the motion of vortices. He did experiments with smoke rings: vortices of smoke can be made to undergo elastic collisions and exhibit curious modes of vibration. These observations led William Thompson (later Lord Kelvin), a close associate of Tait, to propose his vortex model of matter: an atom is a knotted vortex in the ether. This theory was investigated seriously for about 20 years with physicists such as James Clerk Maxwell stating that it had advantages over other contemporary models. The particular point of interest to us is that the atomic vortices of different elements would be knotted in different ways. It is this that led Tait to start his study of knots.

Although the vortex theory was eventually dropped by physicists, Tait continued his investigation. The combinatorial explosion of possibilities for higher orders made progress slow. In 1885, Thomas Kirkman published tables of knot projections (4-valent graphs) [42, 43]. Taking Kirkman's projections, Tait replaced the vertices with crossings to produce knot diagrams. He chose the crossings so that, on travelling along the knot, one goes over then under alternately until returning to the starting point. Such diagrams are called *alternating*. Before publication of his results, Tait learned of another catalogue of knots up to order 10 produced by Charles Little [47]. Comparing the two lists, Tait found one duplication in his own table, and one duplication and an omission in Little's.

Tait encouraged Little to enumerate the alternating diagrams of order 11 starting from Kirkman's graphs. Little also started to consider the possibility of non-alternating diagrams. This dramatically increases the number of cases to be studied. In fact, it took him six years to produce his catalogue of 43 non-alternating diagrams of order 10. Incredibly, his list had no omissions, and only one duplication, which went unnoticed until 1974. By 1900, Tait, Kirkman, and Little had laboured for many years and created catalogues of knots up to order 10, and alternating knots of order 11.

It must be remembered that these catalogues were compiled by hand, with no means apart from experiment to group diagrams into equivalence classes. There are, in fact, two problems that need to be addressed when compiling a catalogue: it must be complete (no omissions), and contain no duplicates. The first of these is an algorithmic problem and is, from a mathematical viewpoint, rather trivial. Proving that knots are distinct is much more difficult, and attention now turned away from enumeration to the development of knot invariants.

The first proof of the existence of non-trivial knots was given by Heinrich Tietze in 1908 [61]. In 1914, Max Dehn showed that the left- and right-handed trefoils are distinct [30]. All the knots up to order 9, except for three pairs, were shown to be distinct in 1927 by James Alexander and his student Garland Briggs [23]. The remaining pairs were separated by Kurt Reidemeister in 1932 [1]. Many techniques, algebraic topology in particular, were developed over time and, as they became more powerful, more and more knots were able to be distinguished. Eventually the catalogues were proved to be remarkably accurate: the tables up to order 10 contained only one duplication.

In the late 1960s John Conway started developing new ideas for extending the knot catalogue. He introduced knot fragments called tangles, which we shall study later, and developed a notation for describing knots in terms of tangles and the ways they can be put together. His powerful notation dramatically simplified the process of enumeration, allowing Conway to verify the existing tables 'in an afternoon'. He went on to enumerate knots up to order 11, and links up to order 10, finding one duplication and 11 omissions in Little's table of alternating knots of order 11. But he still overlooked the duplicate of order 10. In the late 1970s Alain Caudron repeated the enumeration for order 11 and found four omissions in Conway's list.

This marks the end of the era of hand calculation. In the early 1980s, Hugh Dowker and Morwen Thistlethwaite computerised the processes of listing all possible diagrams, dividing them into equivalence classes, and calculating invariants to distinguish them. They were able to extend the catalogue to order 13. In the late 1990s, working with Jim Hoste and Jeffrey Weeks, Thistlethwaite extended the catalogue to order 16 — a list containing over 1.7 million knots [36].

For the first hundred years of knot theory, catalogues contained pictures of unlabelled, unoriented knots and links. Indeed, many catalogues contain only knots. For knots, labelling the components is not an issue. Also, most small knots are reversible, and most invariants of knots do not depend on the orientation chosen. Since the 1970s, the most widely used catalogue has been that in Appendix C of Dale Rolfsen's book [5], compiled by James Bailey and drawn by Ali Roth. It contains almost 400 diagrams of knots and links of small orders.

Calculation of many link invariants requires the link to be oriented; changing the orientations of some components can affect the result, indicating that the links are different. The need for a catalogue of labelled, oriented links was addressed by Helmut Doll and Jim Hoste [31], who also introduced the notation of $+$ and $-$ described in §1.8. The links in their catalogue are taken as the version with all orientations $+$. Some of their diagrams differ from Rolfsen's and depict the mirror-image link (4_1^2 and 6_1^2 in Figure 1.14 for example). The alternating links from Rolfsen's table have been correlated with the Doll–Hoste catalogue by Corinne Cerf [27].

The knot K_i has i sections.

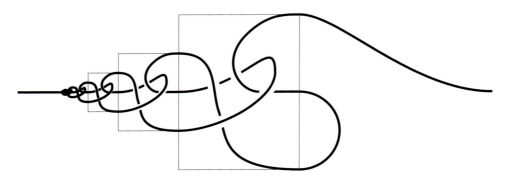

Figure 1.15. The knots K_i are trivial for all $i \in \mathbb{N}_0$ but K_∞ is not.

1.11 A technicality

As we mentioned on page 1, our current definition of a knot suffers from being too general. The following example illustrates the kind of problem that can arise.

An unknotted loop can be deformed, stage by stage, to produce the family of knots shown at the top of Figure 1.15. To construct knot K_i the boxed section is repeated i times; imagine the left and right ends travelling around the back of the book to complete the circle. The knot K_i is defined for all $i \in \mathbb{N}_0$ and each one is trivial. Now consider the knot K_∞ in the lower part of the figure. It is constructed in a similar way except that the sections are scaled by a constant reduction factor and become smaller and smaller, converging to a point. The curve is continuous at the limit point, so it is a knot. It appears as though this knot can be unravelled from the right, just like all the K_i. However, K_∞ is not trivial [35, 34]!

We want to exclude such bizarre behaviour from our study. Clearly, whatever distinguishes K_∞ from each of the other K_i must be a feature of the limit point.

One option to eliminate this wild behaviour would be to require knots to be differentiable: the embedding of K_∞ is not smooth as the derivative is undefined at the limit point. However, this would mean that we could not use polygons, and these are much simpler for many purposes. Alternatively, we could restrict ourselves to using polygons. Since a polygon has only finitely many edges, we cannot construct K_∞ as a polygon and the problem does not arise. Another solution is to introduce some kind of 'local unknottedness' condition. Let $B_p(r)$ denote the closed ball of radius $r > 0$ centred at p: $B_p(r) = \{x \in \mathbb{R}^3 : |p - x| \leqslant r\}$.

Definition 1.11.1 (local flatness).   A point p in a knot K is *locally flat* if there is some neighbourhood $U \ni p$ such that the pair $(U, U \cap K)$ is homeomorphic to the unit ball, $B_0(1)$, plus a diameter. A knot K is locally flat if each point $p \in K$ is locally flat.

The knot K_∞ is not locally flat at the limit point.

A point that is not locally flat is called *wild*, and a knot is wild if any of its points are wild. Unfortunately, this is not a rare phenomenon: almost all knots are wild at every point [25, 53]. However, because wild knots have such counter-intuitive properties, we shall not consider them any further, and revise our initial definition as follows.

Definition 1.11.2 (knot).   A (tame) *knot* $K \subset \mathbb{R}^3$ is a locally flat subset of points homeomorphic to a circle.

Since a link was defined as a collection of knots, our links are tame too.

The examples that we have seen so far using polygons and smooth curves are all locally flat. In fact the three conditions are interchangeable.

Theorem 1.11.3.   *Polygonal knots are locally flat.*

PROOF. Let p be a point in a polygonal knot K. Let E_p be the set of edges of K that do not meet p. If p is not a vertex, p lies in the interior of some edge, say e, and $E_p = K - e$. If p is a vertex then E_p is K minus the two edges which meet at p.

There is always a unique shortest distance between a straight line segment and a point. Because there are finitely many edges in E_p, there is some minimal distance, r, between p and E_p: hence, $B_p(r/2) \cap E_p = \varnothing$.

Furthermore, the ball $B_p(r/2)$ meets K in a piecewise linear curve, so K is locally flat at p. $\square$

Showing that smooth knots are locally flat is more difficult and we shall only give an indication of the method, not a complete proof. Let $f : \mathbb{R} \longrightarrow \mathbb{R}^3$ be a C^1 periodic function that defines a knot. By reparametrising if necessary, we can assume that we travel along it at 'unit speed' so that the tangent vector has length 1 at all points. The curve is then said to be *parametrised by arc length*.

Intuitively, we know that a smooth curve has a well-defined tangent at every point, and if the tangent vector changes direction continuously then we can find a short section of the curve where it does not change direction very much. This argument is made more precise in John Milnor's lemma (see [3, p. 150]):

Lemma 1.11.4. *For any angle θ there exists a $\delta > 0$ such that for any two pairs of points $p_1 < p_2$ and $q_1 < q_2$ in a section of the curve of length less than δ the angle between the vectors $\overrightarrow{p_1p_2}$ and $\overrightarrow{q_1q_2}$ is less than θ.*

Theorem 1.11.5. *A knot which is the image of a C^1 function is locally flat.*

PROOF. Let $f : \mathbb{R} \longrightarrow \mathbb{R}^3$ be the function and let $K = \text{Image}(f)$ be the knot. We shall construct a necklace of conical beads threaded on the knot.

STEP 1. Constructing the cones.

Given two points p and q on the knot, we can measure the distance between them in two ways: we use $d(p, q)$ to denote the Euclidean distance between the points measured in the ambient space $\mathbb{R}^3$; we use $\text{arc}(p, q)$ to denote the shortest distance travelled along the knot.

Consider the map

$$(p, q) \longmapsto \begin{cases} \dfrac{d(p, q)}{\text{arc}(p, q)} & \text{if } p \neq q, \\ 1 & \text{if } p = q. \end{cases}$$

This function is continuous (Exercise 1.12.9) and is defined on a compact set (the knot); therefore it has a minimum value M. The function is positive everywhere so $M > 0$.

Each bead of the necklace is a double cone (see Figure 1.16), which is completely determined by its axial length and the cone angle. (We shall shorten 'double cone' to 'cone'.) The discs orthogonal to the axis are called

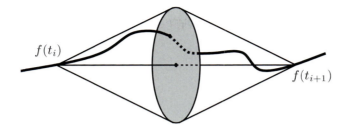

Figure 1.16. A conical bead threaded onto a knot. The angle at each apex is 2θ.

the level discs of the cone. To fix the dimensions we choose an angle $\theta < 45°$ such that $\tan(\theta) < 1/2\,M$. From Milnor's lemma we obtain a δ depending on θ and we divide the total length of the knot into n arcs of equal length λ such that $\lambda \leqslant 1/2\,\delta$. Let $t_i \in \mathbb{R}$ be the points that map onto the endpoints of these arcs. For $1 \leqslant i \leqslant n$ define C_i to be the (double) cone whose axis connects the points $f(t_i)$ and $f(t_{i+1})$ and whose apex angle is 2θ. Unless the knot follows the axis, the axial length will be less than $\mathrm{arc}(f(t_i), f(t_{i+1})) = \lambda$; the cones will vary in size.

STEP 2. The point $f(t)$ lies in some cone for all t.

Let $p = f(t)$ for some $t \in (t_i, t_{i+1})$. By Milnor's lemma, the angle that the axis of cone C_i makes with the vector from $f(t_i)$ to p is less than θ. Similarly for the vector from $f(t_{i+1})$ to p. Hence $p \in C_i$.

STEP 3. Adjacent cones meet only at an apex.

The arc length from $f(t_i)$ to $f(t_{i+2})$ is $2\lambda \leqslant \delta$. Hence, by Milnor's lemma, the angle between adjacent cone axes is at most θ. As the angle at the apex of each cone is $2\theta < 90°$, the cones cannot meet except at their common apex.

STEP 4. Non-adjacent cones do not meet.

Suppose that x is a point that lies in cone C_i. By the continuity of f, the arc of K from $f(t_i)$ to $f(t_{i+1})$ meets every level disc of C_i in at least one point. Let p be a point on this arc that lies in the same level disc as x.

Suppose further that x also lies in cone C_j not adjacent to C_i, and let q be a point on the arc through C_j that lies in the same level disc as x.

The largest distance in a level disc is the diameter of the central disc so $d(p, x) \leqslant \lambda \tan(\theta)$. The cones are not adjacent so $\lambda \leqslant \mathrm{arc}(p, q)$. By construction, $\tan(\theta) < {}^1\!/_2\,M$. Hence

$$d(p,q) \ \leqslant \ d(p,x) + d(x,q) \ \leqslant \ 2\,\lambda\,\tan(\theta) \ < \ \operatorname{arc}(p,q)\,M.$$

But this contradicts the choice of M.

STEP 5. Any level disc in any cone meets the knot exactly once.

That a level disc meets the knot at least once follows from the continuity of f. Suppose that points p and q on the knot lie in the same level disc. Then the angle between the cone axis and the vector $\vec{pq}$ is $90°$, contradicting Milnor's lemma.

STEP 6. Conclusion.

To show that a point $p \in K$ is locally flat we perform this construction, ensuring that p does not lie at the apex of a cone. The double cone containing p is homeomorphic to a ball, and the knot intersects the cone in a single unknotted arc. Therefore a bead plus its thread is homeomorphic to the unit ball plus a diameter. As p can be any point in K, the knot is locally flat.   □

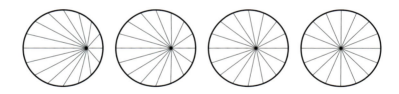

Figure 1.17. Stages in an isotopy of a level disc.

We can extend this proof to obtain the following corollary.

Theorem 1.11.6. *A knot which is the image of a C^1 function is ambient isotopic to a polygonal knot.*

PROOF (SKETCH). We shall describe an isotopy that deforms the knot so that it becomes the set of cone axes. Outside the cones, the isotopy is the identity. Inside each cone, the isotopy behaves on each level disc as indicated in Figure 1.17: the disc is transformed so that the single point of intersection of the knot with the disc is pulled into the centre.   □

The converse is much easier.

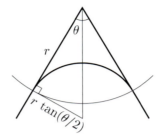

Figure 1.18. A corner can be replaced by a circular arc.

Theorem 1.11.7. *A polygonal knot is ambient isotopic to the image of a C^1 function.*

PROOF. Let e_1 and e_2 be adjacent edges in the polygonal knot K and denote their common vertex by v. Since a polygonal knot is locally flat, there is some $r > 0$ such that the ball $B_v(r)$ meets only two edges of the knot: $B_v(r) \cap K = B_v(r) \cap (e_1 \cup e_2)$. Therefore we can deform the path of K within this ball without changing the knot type of K.

We round off the vertex v by replacing the path of K inside the ball by the arc of a circle that is tangent to the two edges. If the angle between e_1 and e_2 is θ, an arc of radius $r \tan(1/2\, \theta)$ will suffice (see Figure 1.18). This deformation can be achieved by an ambient isotopy which is the identity outside the ball. In fact, K slides in the plane spanned by e_1 and e_2.

As there are finitely many vertices, we can choose disjoint balls, one around each vertex. When the rounding operation is performed at every vertex the resulting knot is the image of a C^1 function. $\square$

This theorem can be generalised to C^∞ functions (Exercise 1.12.11).

1.12 Exercises

1. The picture of the figure-8 knot in Figure 1.2 has four *crossings*: places where one strand of the knot passes over another. Draw pictures of

  (a) three knots with six crossings,
  (b) three 2-component links with six crossings,
  (c) three 3-component links with six crossings.

Can you simplify any of your pictures to reduce the number of crossings? Do any of the pictures represent the same link?

2. A unoriented link $K_1 \cup K_2$ is called *exchangeable* if there is an ambient isotopy which exchanges the two components. Show that each of the four 2-component links in Figure 1.14 is exchangeable. How many are exchangeable when orientation is taken into account?

3. Show the following equivalences for torus knots.

(a) $T(n,1)$ and $T(1,n)$ are the trivial knot for all $n \in \mathbb{Z}$.
(b) $T(p,q) = T(q,p)$.
(c) $T(p,q) = T(-p,-q) = -T(p,q)$; hence torus knots are reversible.
(d) $T(-p,q)$ and $T(p,q)$ are mirror images.

4. The equation for the (p,q) torus knot given in §1.5 is non-linear in the sine and cosine functions. Use trigonometric identities to rewrite its terms as linear combinations of these to give a Fourier expression for torus knots.

5. Show that the frequencies B_x, B_y and B_z of a Lissajous knot can be chosen to be coprime. That is, the highest common factor of any pair is 1.

6. Suppose K is parametrised as a Lissajous knot. Find expressions for K^* and $-K$.

7. Prove that there are no knots with polygon index 4 or 5.

8. There are variants of the polygon index which impose restrictions on the polygon. If the polygon is modelling a long chain molecule so that the vertices represent atoms and the edges are bonds, there is a limited range of allowable edge lengths and angles between them.

Thread some drinking straws onto a string and investigate the polygon index of knots when all the edges have the same length.

9. Let $f : \mathbb{R} \longrightarrow \mathbb{R}^3$ be a differentiable map that is parametrised by arc length. For the ratio of chord length to arc length, show that

$$\lim_{p \to q} \left(\frac{d(f(p), f(q))}{\mathrm{arc}(p,q)} \right) = 1.$$

10. Fill in the details of the proof of Theorem 1.11.6.

11. Find a C^∞ version of Theorem 1.11.7. (Hint: consider $\exp(-1/x^2)$.)

12. Show that the following condition is not sufficient to exclude wild knots: for each point p in a knot K there is some $r > 0$ such that $B_p(r)$ intersects K in a finite number of arcs.

2 A Topologist's Toolkit

Most mathematics texts start from a specified base of mathematical knowledge, add some new ideas, and build on what has been previously studied to develop a familiar topic to a deeper level, or in a new direction. All the stages in the development of the theory are carefully proved and explained with nothing being assumed or hidden. If this pattern were strictly adhered to, knot theory would not be studied at undergraduate level.

To make the study of knot theory mathematically rigorous requires a fair amount of topological sophistication. We have seen that, if we are not careful, our definitions can lead to unexpected results. Trying to understand and exclude wild behaviour was one of the central questions for topologists in the middle of the twentieth century. Many strange spaces and embeddings were created to provide counterexamples to various conjectures. Eventually this work resulted in a firm foundation for geometric topology.

The fact that these foundations have not been studied should not inhibit progress. We shall proceed as working topologists do and just quote the useful results, safe in the knowledge that they have been established elsewhere. Just because we do not understand how a house is constructed that does not mean we cannot live in it.

The results we need are a set of 'tameness' conditions: they provide the reassurance that the topological world behaves as we expect it to. This means imposing a piecewise linear or differentiable structure on our objects of study. Because it is much easier to describe and manipulate piecewise linear objects, we shall work in that category.

With this in mind, this chapter contains a mixture of definitions, examples, facts, and techniques that we shall need throughout the book. The facts are theorems that we shall not prove. Some of them, like the classification of surfaces [71], you may have seen before. The others are topics for your further studies.

2.1 The 3-sphere

Knots and links live in 3-dimensional space. Normally we think of this space as being modelled by $\mathbb{R}^3$. Occasionally, however, we shall find it more convenient to think of links embedded in a compact space and for this we use the 3-dimensional sphere. Here are some descriptions of S^3.

1. Geometrically, the standard 3-sphere is the set of points in $\mathbb{R}^4$ that are one unit away from the origin: $S^3 = \{x \in \mathbb{R}^4 : |x| = 1\}$.

2. The 3-sphere is often thought of as the *1-point compactification* of $\mathbb{R}^3$. This means that we take $\mathbb{R}^3$ and an additional point denoted by ∞ and give it the following topology: a set U in $\mathbb{R}^3 \cup \{\infty\}$ is open if either U is an open set in $\mathbb{R}^3$, or $U = V \cup \{\infty\}$ where V is the complement of a closed and bounded set in $\mathbb{R}^3$.

   To show that $\mathbb{R}^3 \cup \{\infty\}$ is equivalent to the 3-sphere we use stereographic projection (the 2-dimensional case is shown in Figure 2.1). Let $\mathbb{R}^3 \subset \mathbb{R}^4$ be the subspace obtained by setting the fourth coordinate to -1, and let X be the unit 3-sphere in $\mathbb{R}^4$ minus the point $(0,0,0,1)$. There is a homeomorphism $h : \mathbb{R}^3 \longrightarrow X$: the unique line from $(0,0,0,1)$ to a point $p \in \mathbb{R}^3$ cuts X in a single point q and we define $h(p) = q$. We can extend h to a homeomorphism $\mathbb{R}^3 \cup \{\infty\} \longrightarrow S^3$ by setting $h(\infty) = (0,0,0,1)$. Check h is continuous in the given topology.

   If we are working in S^3, pictures of knots in infinite arcs (like the one

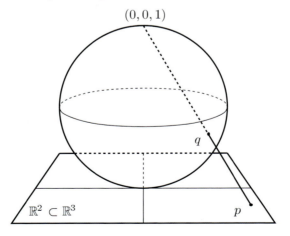

$(0,0,1)$

q

$\mathbb{R}^2 \subset \mathbb{R}^3$ p

Figure 2.1. Stereographic projection of the plane onto a sphere.

shown in Figure 2.2) also represent loops: the lines extending left and right meet at ∞.

3. The (topological) 3-sphere can be constructed as the union of two balls glued together. This is analogous to the constructions in lower dimensions: a circle (S^1) can be formed by gluing the endpoints of two line segments together; a sphere (S^2) is formed by gluing the boundaries of two discs together.

   This decomposition of S^3 is easily seen through the previous example. The lower hemisphere $\{(x_1, \ldots, x_4) \in S^3 \subset \mathbb{R}^4 : x_4 \leqslant 0\}$ is sent by stereographic projection to a 3-ball of radius 2 centred on $(0, 0, 0, -1)$. By symmetry, the upper hemisphere is also (homeomorphic to) a ball.

4. It is possible to construct a piecewise linear 3-sphere. An equilateral triangle is composed of 3 equal line segments meeting 2 at each vertex, and is topologically a circle (S^1). A regular tetrahedron is composed of 4 triangular discs glued edge to edge, meeting 3 at each vertex, and is topologically a sphere (S^2). Proceeding by analogy, we can take 5 solid tetrahedra, glue them face to face with 4 meeting at each vertex: this produces a combinatorial object that is topologically a 3-sphere.

From the topological viewpoint, it does not matter whether we work in $\mathbb{R}^3$ or S^3: a link that passes through ∞ can be moved a little to an isotopic position lying entirely in $\mathbb{R}^3$. However, there are some geometric problems in which differences do show up. In Figure 1.12 we saw that the figure-8 knot can be arranged so that a rigid motion of $\mathbb{R}^3$ (a rotation) carries the knot to its mirror image. In this position, it has an axis of rotation–reflection symmetry where the mirror plane is orthogonal to the axis. For some knots, such a transformation is possible in S^3 but not in $\mathbb{R}^3$. Figure 2.2 shows the oriented knot 8_{17} embedded in S^3 with the line extending left and right to ∞. This oriented knot is not $+$ amphicheiral in S^3 or $\mathbb{R}^3$. However, it is clear from the figure that a 2-fold rotation of S^3 carries the knot to the reverse of its mirror image, showing that it is $-$ amphicheiral. This symmetry can

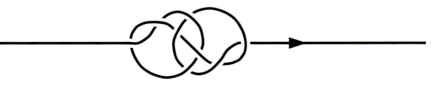

Figure 2.2. The knot 8_{17} is rigidly amphicheiral in S^3 but not in $\mathbb{R}^3$.

be shown topologically in $\mathbb{R}^3$ but it is not possible to demonstrate it using an axis of rotation–reflection in $\mathbb{R}^3$ [21, 70].

2.2 Manifolds

All the topological spaces we are interested in are n-dimensional *manifolds*, usually shortened to n-manifolds. A creature living inside a manifold can not distinguish its environment from the standard space $\mathbb{R}^n$ until it starts to move around. Any neighbourhood looks just like a neighbourhood of $\mathbb{R}^n$; it is only on a global scale that differences between manifolds can be discovered.

The upper half-space in $\mathbb{R}^n$ is the set $\mathbb{R}^n_+ = \{(x_1, \ldots, x_n) \in \mathbb{R}^n : x_n \geqslant 0\}$. Its boundary, the subset given by $x_n = 0$, is a copy of $\mathbb{R}^{n-1}$.

A topological space X is an n-manifold if it is Hausdorff,[*] has a countable basis, and each point $x \in X$ has a neighbourhood $U \ni x$ which is homeomorphic to $\mathbb{R}^n$ or to $\mathbb{R}^n_+$. If $x \in U$ maps to a point in the boundary of $\mathbb{R}^n_+$ then x is a boundary point of X. The set of all boundary points, denoted by ∂X, is an $(n-1)$-manifold and is called the *boundary* of X.

It is clear that a manifold can be covered with open sets, and that the covering may contain uncountably many sets. A manifold X is *compact* if, when given any covering of X by open sets, we can find a finite subcover. A compact manifold with no boundary, that is $\partial X = \varnothing$, is said to be *closed*.

Up to homeomorphism, there are only four different connected 1-manifolds, and an example of each type is given in Figure 2.3. The real line $\mathbb{R}$ itself is clearly a manifold. It has no boundary and is not compact. The half-line has a single boundary component, the point 0. The edge skeleton of a cube is not a 1-manifold: the vertices do not have the required neighbourhoods.

All the 1-manifolds have natural analogues in all other dimensions. The

Example		Boundary	Compact
Line	$\{x \in \mathbb{R}\}$	✗	✗
Half-line	$\{x \in \mathbb{R} : x \geqslant 0\}$	✓	✗
Interval	$\{x \in \mathbb{R} : x^2 \leqslant 1\}$	✓	✓
Circle	$\{(x,y) \in \mathbb{R}^2 : x^2 + y^2 = 1\}$	✗	✓

Figure 2.3. Examples of 1-manifolds and their properties.

[*]A space is Hausdorff if any two points can be placed in disjoint open sets.

line generalises to the plane and 3-space. The half-spaces, $\mathbb{R}^n_+$, are not used much; we are usually more interested in compact spaces. The standard n-dimensional ball is the set $B^n = \{(x_1, \ldots, x_n) \in \mathbb{R}^n : \sum x_i^2 \leqslant 1\}$. A 2-ball is usually called a disc and is sometimes denoted by D^2. The standard n-sphere, S^n, is the boundary of an $(n+1)$-ball. However, by stereographic projection we can also think of S^n as $\mathbb{R}^n \cup \{\infty\}$. Unless it is clear from the context that we are considering a geometrically round ball or sphere, the terms disc, ball, and sphere refer to deformable objects that are homeomorphic to the standard ones.

Our main concern will be with compact manifolds of dimensions 1 and 2. A compact 2-manifold is called a *surface*.

The study of 3-manifolds is a large area of current research. We are interested in only a few examples: $\mathbb{R}^3$, S^3, B^3, the solid torus $D^2 \times S^1$, and manifolds obtained from these by removing neighbourhoods of links. In the latter case, we distinguish the compact manifold, called the *exterior*, from the open one, called the *complement*. So, for example, if $K \subset S^3$ is a knot, then $S^3 - K$ is the knot complement, and S^3 minus a small open solid tubular neighbourhood of K is the knot exterior. The exterior of K, denoted by $\mathrm{ext}(K)$, is compact and its boundary surface is a torus. It is well-defined because K is tame.

2.3 Orientation and orientability

In every dimension there are two choices of orientation: a line segment can be oriented by adding an arrow pointing either forwards or backwards; a patch of surface can be oriented clockwise or counter-clockwise; a region of space can be oriented with a right-handed or left-handed helical screw.

Each point of a manifold can be assigned one of these two orientations, and locally, it is always possible to make all the orientations in a small neighbourhood agree. If it is possible to orient an entire manifold coherently, it is said to be *orientable* and, when a choice of orientation is made, the manifold is said to be *oriented*.

Every 1-manifold is orientable, but in higher dimensions it is not always possible to give a manifold a coherent orientation at every point. The Möbius band is the most well-known example of an *unorientable* surface: it is impossible to establish a consistent notion of clockwise over the whole surface. If the universe is an unorientable 3-manifold, you could travel along an orientation-reversing path. On your return to Earth you would notice your ideas of left and right had been interchanged: writing would be backwards, for example. We shall not be much concerned with unorientable manifolds.

An orientable surface is sometimes called 2-sided: the two sides of the surface can be painted different colours without running into a problem. The two sides are often called positive and negative, and we shall colour them red and black respectively. By convention, the side of a surface on which we see a counter-clockwise orientation is the red or positive side of the surface.

An orientation on a surface determines an orientation on its boundary: take a point close to the boundary and orient the boundary curve so that it 'goes with the flow'. So, for example, if we are looking at a flat disc from above and we see that each point in the disc is oriented clockwise then the boundary as we see it will also run clockwise.

2.4 Separation and bounding

We are interested not only in the intrinsic properties of manifolds but also in the ways that they sit inside one another. The study of knots and links is entirely about this kind of placement problem.

Definition 2.4.1 (embedding). Let $h : X \longrightarrow h(X) \subset Y$ be a homeomorphism. We say that h is an *embedding* and that X is embedded in Y. If X and Y have boundaries and $h(X) \cap \partial Y = h(\partial X)$ then h is a *proper* embedding and X is properly embedded in Y.

Our definition of a knot is equivalent to saying that a knot is a (tame) embedding of S^1 in $\mathbb{R}^3$.

Embedding is a global property of X. We are also interested in spaces that are 'locally embedded'.

Definition 2.4.2 (immersion). We say that a map $h : X \longrightarrow h(X) \subset Y$ is an *immersion* if for all $x \in X$ there is a neighbourhood $U \ni x$ such that $h : U \longrightarrow h(U)$ is an embedding.

For example, a circle can be immersed in $\mathbb{R}^2$ in the form of a figure 8; this is not an embedding because the inverse mapping is not well-defined at the intersection point.

An *arc* is the image of an immersion of the interval $[0, 1]$. An immersed circle is called a *loop*. An embedded arc or loop has no self-intersections and is called *simple*.

Two sets X and Y are *disjoint* if $X \cap Y = \varnothing$. A set Y is *disconnected* if there exist two non-empty disjoint open subsets A and B of Y such that $Y = A \cup B$; if such a partition is impossible then Y is *connected*. Let Y be a connected set and suppose $X \subset Y$. If $Y - X$ is disconnected then X *separates* Y. Loops embedded in a surface, and arcs properly embedded in a surface with boundary, can be either separating or non-separating: we shall often need to discriminate between them.

Fact 2.4.3 (Jordan Curve Theorem). *A loop λ embedded in $\mathbb{R}^2$ separates the plane into two connected components, and each is bounded by λ.*

Fact 2.4.4 (2D Schoenflies Theorem). *If λ is a loop embedded in $\mathbb{R}^2$ then there is a homeomorphism $h : \mathbb{R}^2 \longrightarrow \mathbb{R}^2$ such that $h(\lambda)$ is the unit circle.*

This is one of the most useful theorems in the topology of the plane; it has the following consequences.

- The closure of one component of $\mathbb{R}^2 - \lambda$ is (homeomorphic to) a disc.

- There are no wild knots in the plane.

- All knots in the plane are trivial.

- A loop λ embedded in S^2 separates the sphere into two connected components, and each is a disc bounded by λ.

These two theorems state some basic facts about loops in the plane and, on a first reading, they may seem obvious. However, they are far from easy to prove, and their 'obviousness' evaporates when we realise that one of them generalises to all higher dimensions and the other is false even in three dimensions.

The statement of the Jordan Curve Theorem generalises easily:

Fact 2.4.5. *A sphere S embedded in $\mathbb{R}^3$ separates $\mathbb{R}^3$ into two connected components, and each is bounded by S.*

We saw in Chapter 1 that there are wild embeddings of S^1 in $\mathbb{R}^3$ (see Figure 1.15). This kind of pathology is also possible with embedded spheres. In particular, there are wild embeddings of S^2 such as James Alexander's famous horned sphere [65]. We need to impose some restrictions on our embeddings to exclude such wild behaviour. As before, we can choose local flatness, piecewise linearity, or differentiability: in dimensions up to 3, the topological effect is the same. Throughout the rest of the book we shall assume that all embeddings are tame. With this restriction, we can generalise Schoenflies' Theorem.

Fact 2.4.6 (Schoenflies Theorem).   *If S is a tame embedding of a sphere in $\mathbb{R}^3$ then there is a homeomorphism $h : \mathbb{R}^3 \longrightarrow \mathbb{R}^3$ such that $h(S)$ is the unit sphere.*

Consequently, a (tame) sphere embedded in $\mathbb{R}^3$ separates, and bounds a ball on one side. A (tame) sphere in S^3 bounds a ball on both sides.

We are also interested in tame embeddings of the torus.

Fact 2.4.7.   *A torus embedded in S^3 bounds a solid torus $(D^2 \times S^1)$ on at least one side. A torus standardly embedded in S^3 bounds a solid torus on both sides.*

We shall often make use of these results without referring to them explicitly.

2.5   Innermost and outermost

Let X be a set of disjoint loops embedded in a surface F. If $\lambda \in X$ is a loop which bounds a disc $\Delta \subset F$ such that $\Delta \cap X = \partial\Delta = \lambda$ then λ is called an *innermost loop*, and Δ an *innermost disc*.

Fact 2.5.1.   *A finite set of disjoint loops on a sphere contains at least two innermost loops.*

Let X be a set of disjoint arcs properly embedded in a surface F. If $\alpha \in X$ is a separating arc which cuts off a disc $\Delta \subset F$ such that $\Delta \cap X = \alpha$, and $\partial\Delta = \alpha \cup \beta$ where $\beta \subset \partial F$ then α is called an *outermost arc*, and Δ an *outermost disc*.

Fact 2.5.2.   *A finite set of disjoint arcs properly embedded in a disc contains at least two outermost arcs.*

These two results are stated because they are useful but, unlike most of the other facts in this chapter, they are easy to prove (Exercise 2.13.4).

2.6   Surfaces

A loop is a rather featureless object. As a knot, its only source of interest is as an embedded object in space. A surface, on the other hand, has additional structure which is independent of its embedding. It has intrinsic properties that allow a 2-dimensional creature to differentiate one type of surface from another.

You should already be familiar with several different surfaces: the sphere, torus, disc, and annulus for example. These are all connected orientable surfaces. There are two features that allow us to distinguish them:

Boundary components: The sphere and the torus are closed surfaces (they have no boundary). The disc and annulus do have boundary components: the disc has one, and the annulus has two.

Connectedness: To measure the *connectedness* of a surface we start to cut it open along embedded loops and properly embedded arcs. There is a maximum number of such cuts that can be made before the surface becomes disconnected: the number is a topological invariant of the surface.

On a sphere we cannot make any cuts at all: it has no boundary so there are no properly embedded arcs, and all embedded loops separate it into two pieces. On the torus, a meridian m and a longitude l are two examples of non-separating loops. Furthermore, the set $m \cup l$ is also non-separating: when the torus is cut along both curves, it can be opened out into a rectangular disc. Any set of three or more loops on a torus will disconnect it.

The conclusions of this discussion are summarised in the following table.

	Sphere	Torus	Disc	Annulus
Number of boundary components	0	0	1	2
Number of non-separating loops	0	2	0	0

On an orientable surface, non-separating loops always occur in pairs (see Figure 6.2 on page 138). Hence, an orientable surface always has an even number of such non-separating loops, say $2g$. The number g is called the *genus* of the surface.

The following theorem states that the two numbers $|\partial F|$ and $g(F)$ are sufficient to identify the topological type of a surface.

Fact 2.6.1 (Classification of orientable surfaces). *Two connected orientable surfaces are homeomorphic if and only if they have the same number of boundary components and the same genus.*

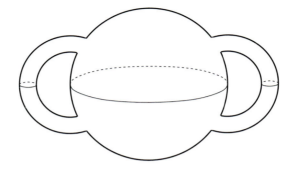

Figure 2.4. A sphere with two handles.

Another way to think of genus is as the number of independent 'tunnels' through the surface or, equivalently, the number of independent 'handles' it has. A *handle* is just a tubular part of a surface. Figure 2.4 shows a sphere with two handles. Figure 6.2 shows that each handle contributes two non-separating curves.

Fact 2.6.2. *A closed connected orientable surface of genus g is homeomorphic to a sphere with g handles attached.*

We can easily make some examples of surfaces that have boundary components by removing discs from the closed surfaces. In fact, all examples can be made in this way.

Fact 2.6.3. *An orientable surface with boundary is homeomorphic to one of the closed surfaces with a set of discs removed.*

Note that the genus of a surface with boundary equals the genus of its associated closed surface.

By expanding the holes, a surface with boundary can be deformed until it looks like a set of ribbons or bands connected together. A deformation of a punctured torus is shown in Figure 5.4. A further step deforms the surface so that it looks like Figure 5.5 — a disc with two bands attached. This can be done with any surface:

Fact 2.6.4. *A connected orientable surface with boundary is homeomorphic to a surface constructed by attaching bands to a disc.*

Two disc–band surfaces are homeomorphic if and only if they have the same number of bands and the same number of boundary components.

We shall explore this further in Chapter 5.

2.7 Euler characteristic and genus

We saw in Chapter 1 that knots can be defined as polygons or as the images of periodic smooth functions. Similarly, surfaces can also be taken to be polyhedral or smooth in nature. However, for our purposes, it turns out to be much easier to use the polyhedral form and consider a surface as a piecewise linear object built from triangles.

Fact 2.7.1. *An orientable surface can be triangulated.*

This means that a surface can always be represented as a finite collection of triangular discs glued along their edges.

When a surface F has been triangulated, its Euler characteristic $\chi(F)$ is defined as

$$\chi(F) \;=\; V - E + T$$

where V, E and T are the numbers of vertices, edges and triangles in the triangulation, respectively.

Fact 2.7.2. *The Euler characteristic is independent of the triangulation and hence is a topological invariant of the surface.*

For a connected surface F, the Euler characteristic is related to the genus as follows:

$$2\,g(F) \;=\; 2 - \chi(F) - |\partial F|$$

where $|\partial F|$ is the number of boundary components of F.

As noted above, a connected orientable surface with n boundary components is homeomorphic to one of the closed orientable surfaces with n discs removed, and its genus is the same as the genus of its associated closed surface. We define the genus of a disconnected surface to be the sum of the genera of its component pieces: $g(F_1 \cup F_2) = g(F_1) + g(F_2)$.

Fact 2.7.3. *Let F_1 and F_2 be two connected orientable surfaces. Let $F_1 \sqcup F_2$ denote their distant union; $F_1 \cup_\alpha F_2$ the union such that $F_1 \cap F_2$ is a single arc in the boundary of each F_i; and $F_1 \cup_\lambda F_2$ the union such that $F_1 \cap F_2$ is a single loop in the boundary of each F_i. Then their Euler characteristics and genera behave as follows.*

F	$\chi(F)$	$g(F)$
$F_1 \sqcup F_2$	$\chi(F_1) + \chi(F_2)$	$g(F_1) + g(F_2)$
$F_1 \cup_\alpha F_2$	$\chi(F_1) + \chi(F_2) - 1$	$g(F_1) + g(F_2)$
$F_1 \cup_\lambda F_2$	$\chi(F_1) + \chi(F_2)$	$g(F_1) + g(F_2)$

2.8 Surgery

Figure 2.4 shows a sphere with two handles. We can attach extra handles to the sphere, or any other surface, as follows. Choose two discs Δ_1 and Δ_2 in a (possibly disconnected) surface F and take a tube $A = S^1 \times [0, 1]$. Then replace the discs with the tube: remove Δ_1 and Δ_2 from F and identify the ends of the tube with the circles $\partial\Delta_1$ and $\partial\Delta_2$. If F is orientable we can add the tube so that the result is also orientable. If both ends of the tube are on the same component of the surface, we have attached a handle; if the ends are on disconnected components then we say they have been *piped together*. This is an abstract description. When the surface is embedded, the tube must be embedded so that only its boundary circles meet F. There are many ways to embed the tube (it could be knotted) and non-isotopic surfaces can be produced. This process and its reverse (described below) are examples of surgery.

Surgery can be performed on any manifold but we are interested only in its application to surfaces. For reasons that will become apparent, the technique is also called 'cut and paste'. Almost all the surgery we shall do will make use of discs in the surface complement, so we shall study this as an example.

Suppose that $F \subset M$ is an orientable surface embedded in a 3-manifold where M is $\mathbb{R}^3$ or S^3. Let $\Delta \subset M$ be a disc such that $\Delta \cap F = \partial\Delta$. This means Δ is properly embedded in the exterior of F. The boundary of the disc, $\partial\Delta$, is a loop in F. Surgery consists of two procedures (see Figure 2.5):

Cutting: We cut F along the loop $\partial\Delta$. Equivalently, we can delete an annular neighbourhood of $\partial\Delta$ from F. This leaves a (possibly disconnected) surface with two new boundary components.

Pasting: We now attach a copy of Δ to each boundary component, producing a new surface $\widehat{F}$.

Such a surgery on a surface can have various consequences.

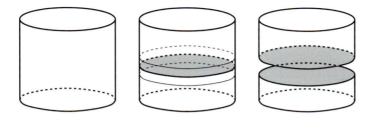

Figure 2.5. Surgery on a hollow tube. The central picture shows a disc Δ properly embedded in the complement of the tube, and a bicollar of $\partial\Delta$ in the tube. On the right the bicollar is removed and two copies of Δ are attached.

- If $\partial\Delta$ is a separating curve in F then $\widehat{F}$ is disconnected. When this happens, it is often the case that one component of $\widehat{F}$ contains all the parts of F that we are interested in (such as the boundary) and the other component is a closed surface. Since it is of no interest, the closed component is simply discarded.

  If the discarded component is a sphere then it bounds a ball, B. If there are no obstructions (such as link components) inside B then the effect of surgery and discarding the sphere can also be produced by an ambient isotopy of F: the disc $\partial B - \Delta$ in F is isotoped across B so that it reaches Δ.

- If $\partial\Delta$ is a non-separating curve in F then this surgery reduces the genus: $g(\widehat{F}) < g(F)$.

Surgery can be performed on F using any orientable surface S properly embedded in the exterior of F. Since S is orientable, there is a map $b : S \times [-1, 1] \longrightarrow M$ such that $b(S \times \{0\}) = S$ and $b(\partial S \times [-1, 1]) \subset F$. The product neighbourhood is called a *bicollar* of S. The surgery consists of removing the set of tubes $\partial S \times [-1, 1]$ and pasting in two parallel copies of S so that F becomes

$$\Big(F - b(\partial S \times [-1, 1])\Big) \ \cup \ b(S \times \{-1\}) \ \cup \ b(S \times \{1\}).$$

We shall see an example of this in step 2 of the proof of Theorem 5.8.1 where S is an annulus.

2.9 Compressibility

If a loop λ embedded in a surface F does not bound a disc in F it is called *essential*.

Definition 2.9.1. A surface F embedded in a 3-manifold M is *compressible* if any one of the following is satisfied:

1. F is a 2-sphere and it bounds a 3-ball in M;

2. F is a disc in ∂M;

3. F is a disc properly embedded in M and there is a 3-ball in M whose boundary is contained in $F \cup \partial M$;

4. F is not a 2-sphere or a disc, and there is a disc $\Delta \subset M$ such that $\Delta \cap F = \partial \Delta$ and $\partial \Delta$ is an essential loop in F. The disc Δ is called a *compressing disc*.

A surface that is not compressible in M is called *incompressible*.

If every 2-sphere in M is compressible then M is said to be *irreducible*. Fortunately, all the 3-manifolds we are interested in are irreducible: $\mathbb{R}^3$, S^3, B^3, the solid torus, and knot complements. With link complements we have to be more careful: the complement of the 2-component trivial link is not irreducible. For this reason, we sometimes specify that links must be non-split to ensure that every sphere bounds a ball.

In general, the surfaces we shall study are embedded in $\mathbb{R}^3$ or S^3 so the second and third cases of the definition do not apply because these 3-manifolds have no boundary.

The fourth condition is more interesting. In this case a compressible surface can be simplified by performing surgery using the compressing disc. The surgery increases the Euler characteristic of F by two: there are two extra faces. If the surface remains connected then the genus is reduced by one, otherwise it stays unchanged.

Note that a surface may have minimal genus and still be compressible. For example, the 2-component trivial link can be spanned by an annulus or by two disjoint discs. Both a disc and an annulus are subsets of the sphere and so both have genus zero. However, the annulus is compressible. For this reason we may sometimes require that surfaces be incompressible rather than minimal genus.

2.10   Transverse intersections and general position

We are concerned not only with individual curves and surfaces, but also with collections of such objects. We need to consider the ways that the members of the collection are related to one another and, in particular, the ways that they intersect. Surfaces in space can intersect in many ways. However, there is a theory that allows us to assume that all intersections have a simple form: it is called the Principle of General Position.

A structure or property is said to be *stable* if it maintains its essential characteristics when it is perturbed a little. We want our intersections to be stable. For example, the parabola $y = x^2$ and the x-axis have a single point of intersection at the origin. However, if we perturb the parabola to be $y = x^2 + \varepsilon$ then the intersection disappears if $\varepsilon > 0$ or becomes two points when $\varepsilon < 0$. So this type of intersection is not stable.

An intersection is said to be *transverse* if a neighbourhood of it is homeomorphic to one of the following four archetypal cases illustrated in Figure 2.6:

- x-axis $\cup$ y-axis in $\mathbb{R}^2$,
- z-axis $\cup$ xy-plane in $\mathbb{R}^3$,
- xz-plane $\cup$ yz-plane in $\mathbb{R}^3$,
- xy-plane $\cup$ xz-plane $\cup$ yz-plane in $\mathbb{R}^3$.

These represent the only stable arrangements of objects in two and three dimensions. A set of objects is said to be in *general position* if all their intersections are transverse.

Fact 2.10.1. *Every finite set of curves and surfaces embedded in $\mathbb{R}^3$ is arbitrarily close to an ambient isotopic set in general position.*

This means that whatever situation we are presented with, we can perturb it a bit so that all its elements are in general position with respect to one another. We have eliminated points of tangency such as saddle points, maxima and minima. Any random finite collection of curves and surfaces is very likely to be in general position.

Two closed surfaces in general position intersect in a set of simple loops. For a surface with boundary and a closed surface in general position, the intersection consists of simple loops and simple arcs; the arcs are properly embedded in the surface with boundary.

2.11 Graphs

A *graph* is a finite set of points V, called *vertices*, and a finite set E of *edges*. Each edge can be represented as an unordered pair $[v_i, v_j] \in V \times V$.

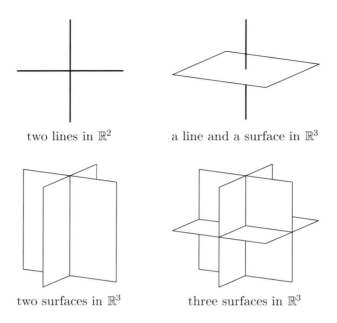

two lines in $\mathbb{R}^2$ a line and a surface in $\mathbb{R}^3$

two surfaces in $\mathbb{R}^3$ three surfaces in $\mathbb{R}^3$

Figure 2.6. Canonical configurations for transverse intersections.

The vertices v_i and v_j are called the *endpoints* of the edge. An edge and its endpoints are said to be *incident* to each other.

Graphs are not manifolds; they are examples of a more general type of object called a *complex*. Their basic properties are probably familiar but we shall review some definitions to establish our terminology.

We do not exclude the possibility that two or more edges share the same endpoints, in which case we have a *multiple edge*. An edge with the same vertex at both ends is called a *loop*. If a graph does not have loops or multiple edges, it is called *simple*: in this case we can regard E as a subset of $V \times V$.

The *valence* (or degree) of a vertex v is the number of times the vertex appears in the list of unordered pairs, E. This is the same as counting the number of edges that have an endpoint at v, with loops counted twice.

Vertices are *adjacent* if they are connected by an edge.

A *path* in a graph is a sequence of edges that can be written as $[v_0, v_1]$, $[v_1, v_2]$, $[v_2, v_3]$, ..., $[v_{n-1}, v_n]$. A path is *simple* if all the v_i are distinct.

A *circuit* in a graph is a path which has the same vertex at the two ends: $v_0 = v_n$. A circuit is *simple* if all the v_i are distinct except that $v_0 = v_n$.

A graph is *connected* if there is a path between any two vertices. A *cut vertex* (or articulation point) is a vertex whose removal disconnects the graph. An *isthmus* is an edge whose removal disconnects the graph.

A *tree* is a connected graph with no circuits. A tree may have one special vertex called the *root*, in which case it is sometimes called a rooted tree. The vertices of a tree are sometimes called *nodes*. The 1-valent vertices in a tree are called *terminal nodes*, and the edges attached to them are called *leaves*.

Fact 2.11.1. *A tree with n vertices has $n - 1$ edges.*

A graph $G' = (V', E')$ is a *subgraph* of a graph $G = (V, E)$ if $V' \subset V$ and $E' \subset E$. Since G' is a graph, the endpoints of all the edges in E' must belong to V'.

A subgraph G' of G is *maximal* with respect to some property P if G' has property P but no other subgraph containing G' has property P. This means that we cannot add extra vertices or edges from G to G' without destroying the property we are interested in. Note that this is a local maximum, not a global one.

A *spanning tree* is a maximal subgraph that is a tree.

A *component* is a maximal subgraph that is connected.

A *block* is a maximal subgraph with no cut vertices.

The *rank* of a connected graph G is the number of edges in G minus the number of edges in a spanning tree. This is a measure of the number of independent circuits in G.

The edges or vertices of a graph can be labelled. A *signed* graph is one in which every edge is assigned a label $+$ or $-$.

The graphs discussed so far exist as abstract structures but we are mainly concerned with embedded graphs. A graph that is embedded in a plane is called a *plane* graph. An abstract graph that has such an embedding is called a *planar* graph.

A graph embedded in a plane divides it into regions, one of which is unbounded. The bounded regions are called *faces*. Faces are *adjacent* if they have an edge in common. A face F is *star-shaped* if it contains some point, x, such that a straight line from x to any point on the boundary of F lies entirely inside F. The point x is called a *star point*.

Planar graphs have the following nice property [69, 77, 79].

Fact 2.11.2 (Fáry's Theorem). *A simple planar graph can be embedded in $\mathbb{R}^2$ so that each edge is a straight line.*

PROOF (SKETCH). Take a plane embedding of a planar graph G. We shall show that G is equivalent to a piecewise linear embedding in which each face is star-shaped. The proof proceeds by induction on the number of vertices.

If G has an internal vertex (one that does not touch the unbounded face) then we delete it and its edges from G. The new graph has fewer vertices so, by induction, it can be embedded as a piecewise linear graph with all faces star-shaped. The deleted vertex can be replaced at a star point and the deleted edges replaced as straight lines.

If all the vertices of G lie on the boundary then G can be embedded as a regular polygon plus some diagonals. This also provides a base for the induction. □

2.12   The combinatorial approach

We have seen that precision in topology requires care and subtlety. A common way to deal with this problem is to work in the piecewise linear category. This means that all the objects studied can be built from finite collections of lines and triangles, and combinatorial methods can be used to analyse how things interact. This restriction removes the possibility of wild behaviour. Indeed, tameness is often defined in terms of piecewise linear objects: 'a knot is tame if it is ambient isotopic to a polygon'.

A smooth object can be approximated arbitrarily closely by a piecewise linear object. We have seen this for links, but it is also true for surfaces. By using this correspondence, results in one category can be transferred to the other, and topology in the two categories turns out to be the same. It can also be shown that restricting the general topological category to locally flat embeddings produces an equivalent theory. (In more than three dimensions these equivalences start to fall apart and higher-dimensional knot theory is thus more complicated.) For most of the time, it will not matter whether we think of links as polygons or not. However, on a few occasions, the piecewise linear structure of surfaces will be exploited when we use their triangulated form.

As an example of the combinatorial approach, we shall examine link equivalence. Let L be a polygonal link embedded in $\mathbb{R}^3$, and let Δ be a triangle such that

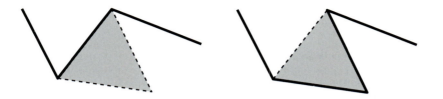

Figure 2.7. A Δ-move changes the path of a polygon.

- L does not meet the interior of Δ,
- L meets one or two sides of $\partial\Delta$,
- the vertices of L in $L \cap \Delta$ are also vertices of Δ,
- the vertices of Δ in $L \cap \Delta$ are also vertices of L.

Define a Δ-move on L as follows: replace L with $(L - (L \cap \Delta)) \cup (\partial\Delta - L)$.

The effect of a Δ-move is illustrated in Figure 2.7. In the degenerate case when Δ has no interior, the move adds a new vertex in the interior of an edge of L, or removes the middle vertex of three consecutive collinear vertices.

Definition 2.12.1 (combinatorial equivalence). Two polygonal links L_1 and L_2 are *combinatorially equivalent* if there is a finite sequence of Δ-moves that transforms L_1 into L_2.

We now have two definitions of equivalence: one based on local moves and one based on global transformation of the ambient space. Clearly a Δ-move can be achieved by an ambient isotopy of the link. Therefore links that are combinatorially equivalent are also ambient isotopic. In fact, the converse is also true.

Fact 2.12.2. *If two polygonal links are equivalent under ambient isotopy then they are also combinatorially equivalent.*

Although we shall not prove this theorem, it lies at the foundation of much of combinatorial knot theory. It gives us an alternative form of the equivalence relation. No longer do we have to analyse a complete 3-dimensional structure — we can study a sequence of much simpler and better controlled situations. This has important consequences and leads to powerful results, as we shall begin to see in Chapter 3.

2.13 Exercises

1. Verify that the piecewise linear object formed from five solid tetrahedra (see page 33) is topologically a 3-sphere.

2. Consider the half-open interval $(-1, 2]$ with the following topology $\mathcal{U}$:

- $\varnothing \in \mathcal{U}$,
- $(a, b) \in \mathcal{U}$ for $-1 \leqslant a < b \leqslant 2$,
- $(a, 0) \cup (b, 2] \in \mathcal{U}$ for $-1 \leqslant a < 0$ and $0 \leqslant b < 2$,
- unions of these.

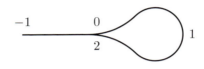

Show that every point has a neighbourhood homeomorphic to $\mathbb{R}$. Show that the space is not Hausdorff and therefore not a manifold.

3. Let F be a connected orientable surface formed by gluing m bands to n discs. Show that $2\,g(F) = 2 - n + m - |\partial F|$.

4. Prove Facts 2.5.1 and 2.5.2.

5. Prove Fact 2.7.3.

6. If a surface consists of two discs and a single band joining them, it is homeomorphic to a single disc with no bands attached. Show by induction that any surface built by adding bands to a collection of discs can be built using only one disc.

7. Fill in the details of the proof of Fact 2.11.2. For example, what happens for 2-valent vertices?

3 Link Diagrams

We have seen several methods for describing links: as polygons with specified coordinates, as the images of smooth functions, but mostly as pictures. Knot theory is definitely a visual subject. In this chapter we investigate 2-dimensional representations of links, their benefits and limitations.

3.1 Pictures of links

Figure 3.1 shows various planar representations of the knot 10_{50}, each with its own pros and cons. Image (a) is quite realistic and appears to show a genuine 3-dimensional object: the shadows and highlights give the impression of depth, allowing the eye to perceive the relative distances of different parts of the knot. It is a computer-generated image produced with the rendering tool POV-Ray. To create it, you need to specify a particular embedding of the knot in $\mathbb{R}^3$, a process which can be time consuming. The result looks rigid and you can get a good idea of how the curve is embedded. The other two images are more schematic with simple conventions to indicate which strands pass over and under each other. The outlined diagram (b) is pleasing and effective for small or simple knots; it was used in Rolfsen's catalogue in [5]. However, it becomes confusing for something complex or with many parallel strands. The simple diagram in (c) is certainly the easiest to draw by hand. When drawn by computer, (b) and (c) are in fact produced in the same way: (b) is a white line with black edges and (c) is a black line with white edges.

3.2 Projections

A very primitive representation of a knot can be formed by projecting it onto a plane. Although a projection gives some information about the embedding in space, it is not sufficient to recreate it: all the height information is lost so we do not know which strings pass over and under. Even so, some projections are still better than others.

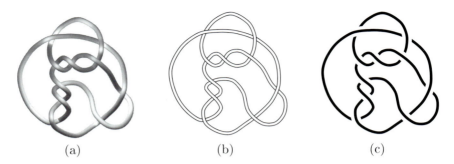

Figure 3.1. Realistic and schematic planar representations of 10_{50}.

Let $L \subset \mathbb{R}^3$ be a link and let $\pi : \mathbb{R}^3 \longrightarrow \mathbb{R}^2$ be a projection map. A point $x \in \pi(L)$ is *regular* if $\pi^{-1}(x)$ is a single point, and is *singular* otherwise. If $|\pi^{-1}(x)| = 2$ then x is called a *double point*.

A projection can be complicated in both the number and the kinds of singular points that occur. Aiming to get a projection that is as clear as possible, we insist that it contains only a finite number of singular points. This means that all the singular points are isolated, and we do not have two edges projecting onto the same line segment, for example. We also want to avoid tangencies, and multiple points of high order where $|\pi^{-1}(x)| \geqslant 3$. We want the projection to be stable so that if we change the direction of projection by a small amount, the projected image is essentially unchanged: no singular points are created or destroyed.

If $\pi(L)$ has a finite number of singular points and they are all transverse double points, the projection is said to be *regular* (see Figure 3.2).

Theorem 3.2.1. *A tame link L has a regular projection.*

PROOF. We noted in §2.12 that, when we are considering tame embeddings, we can choose to work in the piecewise linear category without loss of generality. This allows us to assume that L is a polygon in $\mathbb{R}^3$.

Let $\pi : \mathbb{R}^3 \longrightarrow \mathbb{R}^2$ be a projection. The link can be placed in general position (see §2.10) with respect to the projection. This means that:

- No edge of L is parallel to the direction of projection, hence each edge projects onto a line segment in the plane and not just a point.

- If x is a singular point of the projection then x is a transverse double point, and $\pi^{-1}(x)$ does not contain any vertices of L.

Figure 3.2. A regular projection of some knot — there is insufficient information to say which one.

Any pair of edges will produce at most one intersection point in the projection and there are only finitely many edges in the polygon. Hence there can be only finitely many intersection points in the projection. $\square$

The few examples in Figure 3.3 show that several links can have the same regular projection. A projection with n double points can potentially produce 2^n links since there is a choice of under and over to be made at each double point. However, many of these links will be trivial and other types will also be repeated.

3.3 Diagrams

If a regular projection is annotated in some way so that the observer can tell which strand passes over which at each intersection point then the link type can be recovered. It will not be possible to reconstruct the particular embedding that was projected since the height information is relative, but this does not matter as it is mainly the link type that we are interested in.

Definition 3.3.1 (link diagram). A *diagram* is a regular projection of a link that has relative height information added to it at each of the double points. The convention is to make breaks in the line corresponding to the strand that passes underneath. The double points in the projection become *crossings* in the diagram.

Note that a regular projection has a finite number of double points so a diagram has finitely many crossings. As a simple corollary to the preceding theorem we get

Theorem 3.3.2. *Every tame link has a diagram.*

There are infinitely many different diagrams of every link. This makes

alternating   positive

descending   ascending

Figure 3.3. Different diagrams with the same underlying projection.

trying to identify links by comparing diagrams very difficult. If you make a small knot from a loop of string and try to locate it in the catalogue you will discover this. Simple changes (such as swapping a bounded region with the unbounded region) can produce a diagram that looks very different. It is not easy even to say whether a diagram represents the trivial knot.

In general, there is no 'correct' or 'best' diagram of a link. Indeed, different diagrams may be needed to exhibit different features. The diagrams listed in catalogues are minimal in the sense that they have the smallest possible number of crossings, but such diagrams are not unique. In any case, some link properties may not be apparent in any minimal diagram, as we shall see later.

We shall now define some properties of diagrams. Let $L \subset \mathbb{R}^3$ be a link, $\pi(L)$ be its projection, and D be the diagram constructed from $\pi(L)$. A *subdiagram* of D is a diagram of a sublink $L' \subset L$ using the same projection, so that $\pi(L') \subset \pi(L)$. A *component* of a diagram is a subdiagram corresponding to a 1-component sublink of L.

A diagram is *connected* if its underlying projection is connected. A link that has a disconnected diagram is called a *split* link. Determining when a

Figure 3.4. A diagram with a nugatory crossing can be simplified.

connected diagram represents a split link is another difficult problem: detecting trivial knots and split links are both NP hard in terms of algorithmic complexity [88].

A diagram is *reducible* if the underlying projection has a cut vertex. This means that there is a circle in the plane of the diagram that meets the diagram at a single point which is a crossing (Figure 3.4). Tait called such crossings *nugatory* (meaning unimportant or worthless): they can clearly be removed by untwisting and turning over part of the diagram. A diagram without such crossings is called *irreducible* or *reduced*.

A diagram can also be *oriented*: each component is given an orientation (or inherits one from the link via the projection map). If all the components of a link are oriented, each crossing in a diagram will look locally like one of those in Figure 3.5. These two types of crossing are called positive and negative.

Two forms of link diagram are often used as examples: alternating and positive. A diagram is called *alternating* if one passes over and under alternately as one follows each component (Figure 3.6). An oriented diagram is called *positive* if all its crossings are positive. (In practice, a diagram in which all the crossings have the same sign is also called positive, even if all the crossings are negative.)

A diagram of a knot is called *descending* if it is possible to choose a starting point and a direction so that following around the knot each crossing

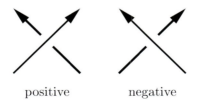

positive   negative

Figure 3.5. Crossings in oriented diagrams come in two types.

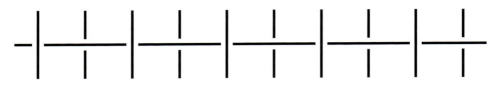

Figure 3.6. The experience of someone travelling along a component of an alternating diagram: passing over and under alternate strands.

is first encountered on the over-crossing strand. An example is shown in Figure 3.3: the arrowhead marks both the basepoint and the direction. For diagrams with more than one component, each component must be given a basepoint and an orientation, and also the components must be ordered $1, \ldots, \mu$. Each component is followed in turn. The complementary concept of an *ascending* diagram is obtained by demanding that each crossing is first encountered on the under-crossing strand.

Families of links can be defined in terms of their diagrams. A link is called *alternating* if it has an alternating diagram. A link is called *positive* if it has a positive diagram. For links with two or more components, positivity can depend on the orientations of the components. The other two types of diagram illustrated in Figure 3.3 do not give rise to new varieties of link: a descending or ascending diagram always represents a trivial link (Exercise 3.10.4).

A diagram is not encumbered with a strong 3-dimensional look. This makes it easy to imagine modifying the diagram by switching over- and under-crossing strands at the crossings. This localised operation is called *switching* a crossing. If crossings are switched selectively then any diagram can be converted into a diagram of a trivial link (Exercise 3.10.4). Hence any projection can be the projection of a trivial link.

3.4 Some families of links

We have already defined a few families of links in terms of some 3-dimensional geometrical properties: torus knots, Lissajous knots, and so on. We can define many more simply by drawing pictures of them.

A sequence of half-twists in a diagram can be replaced by a box containing an integer: the sign* of the integer indicates the sense of the twist as shown in Figure 3.7. Using this shorthand, it is straightforward to draw families of diagrams by connecting these twist-boxes together. Some ex-

---

*There is a choice here; note that other books may use the opposite convention.

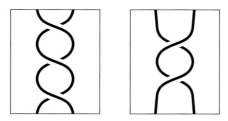

Figure 3.7. Twist-boxes with $+3$ and -2 half-twists.

amples with up to three twist-boxes are shown in Figure 3.8. The single twist-box in (a) closes to the $(p, 2)$ torus link. When $q = 2$, the knot in family (b) is called a *twist knot*. The links in the family (d), denoted by $P(p, q, r)$, are called *pretzel links* since they can be embedded in a closed surface of genus 2, sometimes called a pretzel surface. This is a natural generalisation of the torus knots.

Parts (b) and (c) of Figure 3.8 are examples of rational links. We shall meet them in Chapter 4 and study them in more detail in Chapter 8. The general forms are shown in Figure 3.9: the arrangement depends on whether the number of twist-boxes is odd or even. Such diagrams are denoted by $C(a_1, a_2, \ldots, a_n)$. Note that the twist-boxes in the lower row contain negative signs: this means that the diagram is alternating when all the coefficients a_i have the same sign. The two diagrams in Figure 3.15 show the same knot; they are the rational diagrams $C(4, 1, 5)$ and $C(4, 2, -2, 2, -2, 2)$. Thus the same knot may be represented more than once in a family. It is also possible to represent some links in more than one family. For example, $C(-p, -2) = P(p, 1, 1) = P(p - 1, -3, 1)$. More generally, $C(p, q) = P(-p - 1, 1, q - 1)$. Try to show some of these equivalences (Exercise 3.10.12).

3.5 Crossing number

Since every (tame) link has a diagram (Theorem 3.3.2), we can use link diagrams to define link invariants. Let $c(D)$ denote the number of crossings in a diagram D. The *crossing number* of a link L is the minimum number of crossings in any diagram D of L, and is denoted by $c(L)$:

$$c(L) \;=\; \min\{c(D) : D \text{ is diagram of } L\}.$$

The number $c(L)$ is also called the *order* of L. The diagrams in catalogues of links are arranged in order of increasing crossing number.

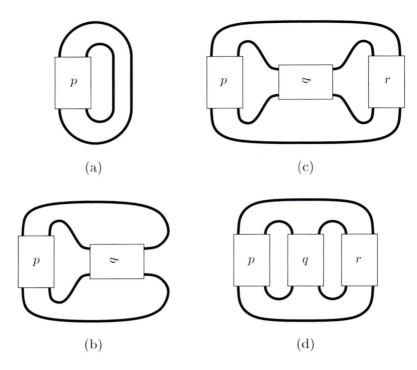

Figure 3.8. Some simple families of links. Each box represents a sequence of half-twists. Diagram (a) is the $(p, 2)$ torus link, (b) is a generalised twist link, and (d) is the (p, q, r) pretzel link.

Theorem 3.5.1. *Let $L \subset \mathbb{R}^3$ be a non-split polygonal link with n edges. If some component of L is non-planar then*

$$c(L) \ \leqslant \ \tfrac{1}{2}(n-1)(n-4).$$

PROOF. We want to calculate the maximum number of double points that can be produced when projecting L onto a plane. An edge cannot give rise to double points with itself or either of its two neighbours, and it can cross each of the other $n - 3$ edges at most once. Hence each edge can contribute at most $n - 3$ double points. Summing up the contributions from all the edges, we shall count each double point twice. Hence there can be at most $\tfrac{1}{2}n(n-3)$ double points in the projection. This bound can be achieved if L is a knot, n is odd, and the projection has a star configuration.

The crossing number is the minimum possible number of double points taken over all possible projections. We can choose to project L along the

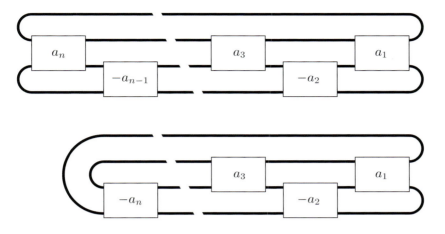

Figure 3.9. Diagrams of rational links denoted by $C(a_1, a_2, \ldots, a_n)$. The format depends on whether n is odd or even.

direction of one of its edges e, in which case this edge becomes invisible in the projection (it is seen end-on and looks like a vertex). The same argument as above applied to the remaining $n - 1$ edges gives the worst case for $c(L)$. Because L has a non-planar component, we can choose e so that the edges adjacent to it project to distinct lines.   □

Corollary 3.5.2. *For a non-trivial knot K*

$$c(K) \ \leqslant \ \tfrac{1}{2}\,(p(K) - 1)(p(K) - 4).$$

PROOF. Choose a minimal polygon for K. A planar knot is trivial, so we know that K is not planar.   □

Inverting the inequality and solving for n gives us a lower bound on the polygon index:

Corollary 3.5.3. *For a link L as in the preceding theorem*

$$p(L) \ \geqslant \ \tfrac{1}{2}\left(5 + \sqrt{8\,c(L) + 9}\right).$$

We can also obtain an upper bound on p from the crossing number [94]. First we need a definition. The projection underlying a diagram D can be thought of as a plane graph G. If two 2-gons in G share a common edge we say that D has *adjacent* 2-*gons*. Part of D will contain an unknotted component with no self-crossings like the ring shown on the left in Figure 3.10.

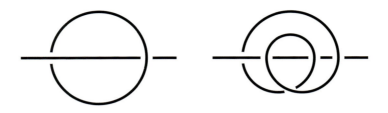

Figure 3.10. Part of a diagram containing adjacent 2-gons (left) and one way to get rid of them.

Theorem 3.5.4. *Let D be a connected irreducible n-crossing diagram of a non-trivial link L. If D does not have adjacent 2-gon faces then L can be embedded as a polygon in $\mathbb{R}^3$ with $2n$ edges.*

PROOF. The diagram D has an underlying projection which we regard as a plane graph G. Each crossing in D becomes a 4-valent vertex in G. Since D is irreducible, G has no loops.

First, let us assume that G is a simple graph. This restriction means that there are no 2-gon faces. By Fact 2.11.2, G is ambient isotopic in the plane to a piecewise linear embedding. So we can assume that G is a graph in which all the edges are straight lines that intersect only at vertices. Each 4-valent vertex can be separated to produce two 2-valent vertices and these can be placed one above the other in the manner needed to produce a crossing of the appropriate type. Figure 3.11 shows this procedure applied to the Borromean rings.

If G is not simple, it must have multiple edges but, as there are no adjacent 2-gons, each multiple edge is a double edge. We can construct a simple graph G' from G by replacing each double edge with a single edge. As before, we can assume that G' is embedded as a straight line graph. We can

Figure 3.11. Constructing a polygonal representation of the Borromean rings. There are two layers of vertices in the far right image; only the top ones are visible.

convert G' back to G by replacing the lost edges: each doubled edge becomes two coincident lines. Now, when we separate the vertices to produce the over- and under-crossings, these doubled edges cause problems. Since they are coincident in G, either they will sit one directly above the other, or they will intersect in a midpoint. Such intersections can be removed by moving one of the endpoints a small distance, taking care to produce the correct link type.

The number of edges in the polygonal representation of L will be the same as the number of edges in G. Since G is 4-valent, this equals twice the number of vertices, or $2\,c(D)$. □

If a link has an unknotted component that links another component like the ring shown on the left in Figure 3.10, the link is said to have the Hopf link as a factor. (We shall meet the notion of link factorisation in Chapter 4.)

Corollary 3.5.5. *If a non-split non-trivial link L is not the Hopf link, and does not have the Hopf link as a factor, then $p(L) \leqslant 2\,c(L)$.*

PROOF. Take a minimal-crossing diagram of L. If the diagram does not contain adjacent 2-gons then the preceding theorem applies and we get $p(L) \leqslant 2\,c(L)$.

When the diagram does contain adjacent 2-gons, it must contain a ring, as shown in Figure 3.10. If both crossings in the ring are the same (both over-crossings or both under-crossings) then the ring lies above or below the other strand and is a split component. In this case it is possible to simplify the diagram of L removing two crossings, contrary to assumption. So the ring must link the other strand as shown in the figure. This means that L is the Hopf link or has the Hopf link as a factor. □

The condition cannot be removed: the Hopf link requires a triangle for each component so $p = 6$ but it has a diagram with only two crossings.

We can also deduce the converse of Theorem 3.3.2:

Theorem 3.5.6. *A link diagram represents a tame link.*

PROOF. If the given diagram contains adjacent 2-gons, we can remove them by adding extra crossings to the diagram as shown in Figure 3.10. The new diagram represents the same link as the original one, and still has finitely many crossings. Applying Theorem 3.5.4 produces a polygonal embedding of the link, hence it is tame. □

In general, crossing number is a difficult invariant to calculate. For many years it had to be found by the exhaustion of possibilities (like polygon index): to prove that an n-crossing diagram is minimal you had to enumerate all the knots or links with fewer than n crossings and then show that your starting link was not in the list. This meant that the only links with known crossing number were the ones in the catalogues. In Chapter 9 we shall see how to compute a lower bound on crossing number from a polynomial link invariant.

3.6 Invariants from oriented diagrams

If a diagram D is oriented, we can form algebraic crossing numbers by summing the signs of various crossings. Let c be a crossing in an oriented diagram D and define

$$\varepsilon(c) = \begin{cases} +1 & \text{if } c \text{ is a positive crossing,} \\ -1 & \text{if } c \text{ is a negative crossing.} \end{cases}$$

If we write '$c \in D$' to mean 'c is a crossing in D' then the *writhe* of D is defined as

$$w(D) = \sum_{c \in D} \varepsilon(c).$$

Just as we defined $c(L)$ from $c(D)$, we can try to use $w(D)$ to obtain an invariant of oriented links:

$$w(L) = \min\{|w(D)| : D \text{ is diagram of } L\}.$$

However, this invariant is ineffectual: $w(L) = 0$ for all L (Exercise 3.10.7).

It is a simple matter to change the writhe of a diagram by adding extra crossings to it. What happens if we restrict attention to diagrams with minimal crossing number? For many years, it was thought that all minimal diagrams of a link had the same writhe. This 'invariant' was used by the creators of catalogues to distinguish links, with the result that the two 10-crossing diagrams shown in Figure 3.12 were classified as distinct knots for 75 years. It was only in 1974 that Kenneth Perko [96] observed that the two diagrams actually represent the same knot, disproving the writhe conjecture.

We shall not give up on algebraic crossing numbers yet. We have seen that crossings of a component with itself lead to problems so we shall exclude them from the calculation.

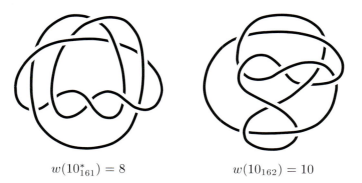

$$w(10^*_{161}) = 8 \qquad\qquad w(10_{162}) = 10$$

Figure 3.12. The famous 'Perko pair' [96]. These two minimal-crossing diagrams of the same knot have different writhes.

Definition 3.6.1 (linking number). Let D be an oriented diagram of a 2-component link $K_1 \cup K_2$, and let D_i denote the component of D corresponding to K_i. The crossings of D are of three types: D_1 with itself, D_2 with itself, and D_1 with D_2. We shall concentrate on the last group which we shall denote by $D_1 \cap D_2$. The *linking number* of D_1 with D_2 is defined to be

$$\mathrm{lk}(D_1, D_2) \;=\; {}^1\!/_2 \sum_{c \in D_1 \cap D_2} \varepsilon(c).$$

We shall prove later (Theorem 3.8.2) that this number is independent of the diagram D and therefore is an invariant of the link $K_1 \cup K_2$. Hence we can write $\mathrm{lk}(K_1, K_2)$.

It is clear from the definition that linking number is a symmetric function: $\mathrm{lk}(K_1, K_2) = \mathrm{lk}(K_2, K_1)$. What this invariant is measuring is the number of times that one component winds around the other. This simple invariant is good enough to provide our first proof that non-trivial links do exist. The linking numbers of the trivial 2-component link, the Hopf link, Solomon's seal, and the Star of David are 0, 1, 2 and 3 respectively (see Figure 1.14). So these four links are all distinct. Unfortunately, the Whitehead link has linking number 0 like the trivial link.

For links with more than two components we can define the *total linking number* to be the sum of the linking numbers of all pairs of components: let $L = K_1 \cup \cdots \cup K_n$ be an n-component link and define

$$\mathrm{lk}(L) \;=\; \sum_{i<j} \mathrm{lk}(K_i, K_j).$$

This allows us to see that 8_4^3 (see Figure 1.14) is non-trivial, but it is not powerful enough to prove that the Borromean rings are truly linked.

3.7   A warning about minimal diagrams

One might naïvely expect that minimal-crossing diagrams are somehow special. Such optimism is misplaced. We have seen that the writhe provides an example where this expectation turned out to be wrong. We must also guard against the belief that minimal-crossing diagrams are the only thing one needs to study: that if a link has a particular property then it will be visible on a minimal-crossing diagram. The falsehood of this idea will be demonstrated with three examples here. Further examples will be discussed later: minimal-genus projection surfaces on page 105, and minimal bindings for arc presentations on page 266.

Positive diagrams : In the diagram on the left of Figure 3.13, all 12 of the crossings have the same sign, so it depicts a positive knot. On the right of the figure is a minimal-crossing diagram of the same knot. Computer enumeration shows it is essentially unique — any other 11-crossing diagram is a simple rearrangement of this one. However, the circled crossing has a different sign from the others.

Amphicheiral knots : Suppose that D is a diagram of an amphicheiral knot and that the symmetry property is observable in the diagram. Since reflection of D in the plane of projection interchanges positive and negative crossings, for D and D^* to be the same there must be an equal number of each type, and so $w(D) = 0$ and $c(D)$ is an even number. Figure 3.14 shows an amphicheiral knot of order 15: the symmetry can be seen on a non-minimal diagram with 16 crossings.

Figure 3.13. A positive knot (11n183) that does not have a positive minimal-crossing diagram [92, 97].

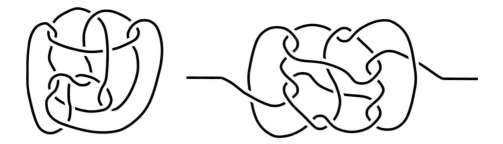

Figure 3.14. An amphicheiral knot (15n139717) whose symmetry is not apparent on a minimal-crossing diagram [36].

Unknotting number : An unknotting operation involves passing one strand of a knot through another strand. This is easily achieved on a diagram by switching the over- and under-crossing strands at a crossing point. The lower diagram in Figure 3.15 is a 14-crossing diagram of the knot 10_8: switching the two circled crossings produces a diagram of the trivial knot. This is the best that can be achieved as $u(10_8) = 2$ (see page 154). The diagram in the upper part of the figure is a minimal-crossing diagram of the same knot, but no two crossing switches will unknot it. Because alternating diagrams have a certain rigidity, the uniqueness of the minimal diagram can be proved without resorting to a computer.

Notice that 'small' knots do not feature in these examples — the simplest ones are 10_8 and 10_{161}. There are only 250 knots of orders up to 10, and

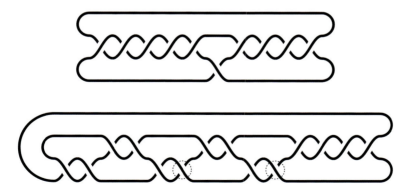

Figure 3.15. A minimal-crossing diagram that cannot be unknotted with the minimum number of crossing switches [82, 93].

about 80% of these are alternating. Until 1970 this small catalogue was the primary dataset for anyone wanting to make or test conjectures. With such a limited amount of data available, it is not surprising that some of the observed patterns of behaviour do not continue to hold for knots in general. A computerised database of the 1.7 million knots of orders up to 16 now exists [254], and this makes it much easier to search for patterns and counterexamples.

3.8 Moves on diagrams

A move is a small localised change to a diagram. Some examples are shown in Figures 3.16 and 3.17. A move is performed on a complete, possibly large, diagram of which only a small part is shown. The broken circle indicates the neighbourhood of activity: no changes are made to the diagram outside the broken circle, and no parts of the diagram except those indicated are contained within the broken circle. A move consists of replacing the diagram fragment in one circle by the one in its partner. Moves often preserve the link type but this is not a requirement.

At the top of Figure 3.16 is a simple move known as a *flype*.* The circle marked 'F' contains a fragment of the diagram and is called a tangle. In the diagram on the right, the tangle has been reflected in a horizontal axis and also in the plane of the page: this is equivalent to turning over the tangle about a horizontal axis.

The lower part of Figure 3.16 shows an example of a *pass move*. As before, the inner circle contains a diagram fragment, but this time it is unchanged by the move. The example shown is just one of a family: there can be any number of strings on the left or the right. If, as in this case, these numbers both equal n then the move is called an n-pass. The example shown is a 5-pass.

It is clear that both flypes and pass moves preserve the link type of the link depicted in the diagram. Tait and Little used flypes to classify alternating diagrams into equivalence classes. They believed that two irreducible alternating diagrams represented the same link if and only if they were related by a sequence of flypes — a conjecture that was proved true a century later by William Menasco and Morwen Thistlethwaite [90, 91]. Little also

---

*This is a Scottish word meaning 'to turn inside out'. Tait used it to denote the operation on a diagram of swapping a bounded region with the unbounded region of the plane — more appropriate than today's usage.

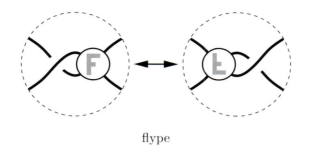

flype

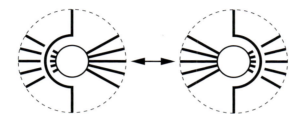

pass move

Figure 3.16. Local changes to link diagrams.

worked on non-alternating diagrams and used flypes and 2-passes to divide them into classes. The Perko pair are a consequence of this: flypes and 2-passes preserve the writhe so they cannot be used to show that these two diagrams represent the same knot.

In the 1920s James Alexander (working in Princeton) and Kurt Reidemeister (in Vienna) independently developed the combinatorial approach to knot theory based on polygons and Δ-moves. This introduced new moves on diagrams. Figure 3.17 shows what are now collectively known as the *Reidemeister moves*. It is clear that they preserve link type. A type 1 move is called adding or removing a *curl*. The type 3 move is often shown as in Figure 3.18. This makes it much clearer what is happening (one strand is being pushed past a crossing involving two other strands) but it destroys the symmetry in the situation and makes the moved strand seem different from the others.

We do not need to include the mirror-image moves in this list. The mirror image of a type 2 move can be obtained by turning the picture upside down, and the mirror images of the others can be synthesised by using sequences of those shown (Exercise 3.10.17).

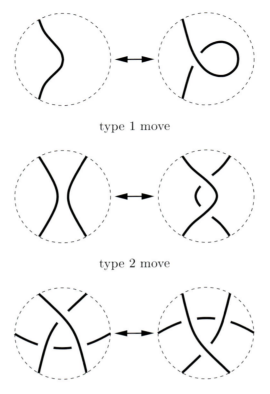

Figure 3.17. The Reidemeister moves.

We now come to what is probably the most fundamental theorem in combinatorial knot theory: these three moves, together with isotopy in the plane, are all that is required to transform a given diagram of a link into any other diagram of the same link.

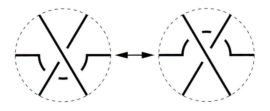

Figure 3.18. An alternative form of the type 3 Reidemeister move.

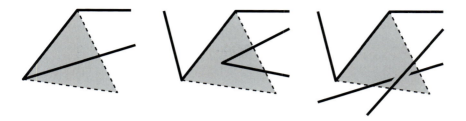

Figure 3.19. Possible projections of a Δ-move.

Theorem 3.8.1. *Two diagrams of a link are related by a finite sequence of Reidemeister moves.*

PROOF (SKETCH). Let D_1 and D_2 be two diagrams of a link L, let $\pi : \mathbb{R}^3 \longrightarrow \mathbb{R}^2$ be a projection, and let L_i be an embedding of L such that $\pi(L_i)$ is mapped onto D_i in the natural way.

We choose to work in the piecewise linear category and assume that our links are polygons $\mathbb{R}^3$. From Fact 2.12.2 we know that any two polygonal embeddings of a link are connected by a finite sequence of Δ-moves. To prove the theorem we need only check what happens to a diagram when one of these moves is performed.

The links can be placed in general position with respect to the projection. This means that

- no edge is parallel to the direction of projection,

- each singular point x of the projected link is a transverse double point, and $\pi^{-1}(x)$ does not contain any vertices of the polygons,

- a triangle used in a Δ-move projects to a triangle in $\mathbb{R}^2$.

Let T be the interior of the projection of the triangle used in a Δ-move. If T does not intersect the diagram then the effect of the Δ-move on the diagram can be obtained by an isotopy in the plane. If T does meet the diagram then, by subdividing the triangle into many small triangles and factorising the Δ-move into a sequence of smaller moves if necessary, we can assume that T contains at most one crossing of the diagram. It remains to analyse the possibilities and to show that each Δ-move corresponds to a combination of Reidemeister moves on the diagram: some representative cases are shown in Figure 3.19. $\qquad\square$

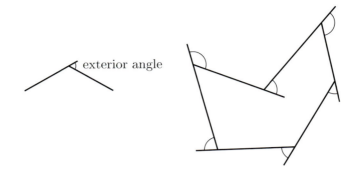

Figure 3.20. Winding number depends on the sum of the exterior angles.

In practice, one never sets about discovering a sequence of Reidemeister moves to prove equality of diagrams. The power of this theorem stems from its capacity to produce link invariants. Suppose that we have a function defined on link diagrams. If its value stays the same when the diagram is changed by any Reidemeister move then the function is independent of the choice of diagram, and is therefore an invariant of the link depicted in the diagram.

As an immediate corollary, we get our first computable link invariant.

Theorem 3.8.2. *The linking number defined above (Definition 3.6.1) is an invariant of oriented links.*

Proof. The computation of linking number does not involve crossings of a component with itself so type 1 moves will not affect its value.

If the two strands in a type 2 move are from different components then the move will introduce or remove two crossings of opposite signs; again the outcome of the computation is unaffected.

A type 3 move changes only the relative positions of crossings; it does not alter their signs or which components of the link are involved. Hence, this move also leaves the sum unchanged. □

Showing the invariance of a function under the Reidemeister moves is a common way to establish that the function is a link invariant, and we shall use it many times.

Definition 3.8.3 (winding number). Let D be a diagram of an oriented polygonal knot. Each vertex in the diagram has an exterior angle (see

Writhe :	+1	+1	−1	−1
Winding number :	+1	−1	+1	−1

Figure 3.21. The four classes of curl.

Figure 3.20): it is the angle you must turn through to continue a walk along the diagram. The angle is signed: as usual counter-clockwise is positive. The sum of these exterior angles divided by $360°$ is called the *winding number* of D. The winding number of a link diagram is the sum of the winding numbers of its components.

The simple diagram on the right of Figure 3.20 has winding number -1. The definition can be extended in a natural way to smooth diagrams by using tangent vectors to monitor the changes in direction.

Like writhe, the winding number depends on the diagram chosen and is not a link invariant. However, the two taken together can provide useful information. Figure 3.21 shows that four types of curl can be distinguished by their contributions to the writhe and winding number. This highlights an important geometrical difference between type 1 Reidemeister moves and the other two [98]:

Theorem 3.8.4. *Two diagrams of an oriented link are related by a finite sequence of type 2 and type 3 Reidemeister moves (without the use of type 1 moves) if and only if they have the same writhe and winding number.*

Proof (sketch). If two diagrams are related by a sequence of type 2 and type 3 Reidemeister moves, they must have the same writhe and winding number since these values are not affected.

To prove the converse, let D_1 and D_2 be two diagrams of a link which have the same writhe and winding number. There exists a finite sequence of Reidemeister moves that transforms D_1 into D_2. We replace the use of type 1 moves in the sequence with the following procedure.

We use the 'Whitney trick' (see Figure 3.22) to insert two curls: the one we want (for the type 1 move) and a dual curl to preserve the writhe and winding number. The important thing is that the addition of two dual curls can be achieved using only type 2 and type 3 moves. A curl can be passed through a crossing using only type 2 and type 3 moves so the additional

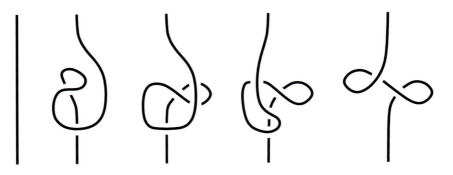

Figure 3.22. The Whitney trick for inserting or removing pairs of curls.

curl can be moved out of the way and we can continue with the next move in the sequence.

We now have a sequence from D_1 to a diagram $\widehat{D}_2$ that looks like D_2 with extra curls. The extra curls can be transported around the diagram $\widehat{D}_2$ and collected together. Since D_2 and $\widehat{D}_2$ have the same writhe and winding number, the curls must occur in dual pairs and the Whitney trick can be used to cancel them, leaving D_2. $\square$

We have seen that the three Reidemeister moves are sufficient to change a diagram into any other diagram of the same link. They are also necessary. Writhe is a property of diagrams that is changed by type 1 moves but not by moves of types 2 and 3. Therefore we cannot produce the effect of a type 1 move by some clever combination of the other two moves. It is possible to define other properties of diagrams that are changed only under type 2 moves, or only under type 3 moves [80, 95]. This shows that all three moves are independent.

3.9   The classification problem

One of the primary goals in any branch of mathematics is to classify the objects of study. A classification is a list of objects known to contain all possibilities without repetition. The usefulness of such a list depends on the criteria used in the classification and how closely they are related to structural properties of the objects. For example, the classification of surfaces up to homeomorphism type reduces to knowing three things: the Euler characteristic, orientability, and the number of boundary components.

It is often straightforward to produce an algorithm that will list all possible objects. The trouble is that the list will contain duplicates and the problem then becomes one of detecting equivalence, which is usually much more difficult. The early knot tabulators could systematically produce lists of knot diagrams with increasing numbers of crossings. Their problem was to identify which diagrams represented the same knot. Tait and Little used flypes and 2-passes to divide diagrams into equivalence classes and the first catalogues were derived by experiment. The tabulators were aware of the provisional nature of their results but they had no invariants to prove that the knot types really were distinct.

The knot equivalence problem (are these two knots the same?) and the knot triviality problem (is this knot trivial?) are examples of decision problems — questions with yes/no answers. For many decision problems it is easy to find an algorithm that will eventually output 'yes' in the case of a positive solution, but will never terminate if a solution cannot be found. With these partial solutions we cannot tell whether failure to finish indicates that a positive answer does not exist or only that we did not wait long enough to find it. To provide a complete solution, an algorithm must terminate in a finite time when given any valid input, and it must output 'yes' or 'no'.

In the early 1930s Kurt Gödel proved his famous incompleteness theorem: any logical system of sufficient complexity contains statements that cannot be proved or disproved using the axioms of the system. The basic rules of arithmetic form a sufficiently complex system so mathematics contains unsolvable problems: statements whose truth or falsehood cannot be determined. Alan Turing adapted Gödel's method to computational models and proved that some decision problems cannot be solved by algorithmic means. Showing that a problem is unsolvable does not mean that we need to try harder: it says that it is impossible to find an algorithm that solves the problem for all valid inputs. One might hope that unsolvability afflicts only artificially contrived problems but, unfortunately, many natural questions have been shown to be unsolvable. The equivalence problem for 4-manifolds is one example.

It was also at about this time that the combinatorial approach to knot theory, based on finite polygons and Δ-moves, was developed. Combinatorial decision problems were some of the first to be proved unsolvable, and people asked whether the knot equivalence and knot triviality problems were solvable.

Can the Reidemeister moves be used to decide when two knots are equivalent? Or, as a special case, can we use them to determine whether a given diagram represents the trivial knot? The second question appears straightforward: find an algorithm that applies Reidemeister moves, simplifying the

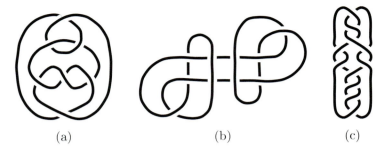

$$(a) \qquad\qquad\qquad (b) \qquad\qquad\qquad (c)$$

Figure 3.23. Awkward diagrams of the trivial knot: the only Reidemeister moves that can be performed increase the number of crossings.

diagram at each stage, and continue until it has no crossings; if no simplification is possible then the diagram cannot be trivial. However, Figure 3.23 shows diagrams of the trivial knot that play havoc with this approach: the only Reidemeister moves that can be applied in these cases increase the number of crossings. Example (c) was created by Lebrecht Goeritz in 1934 [158]; he tried adding pass moves to the list of allowed moves, but found another diagram which could only be simplified after an increase in crossing number. It is possible that this will happen however many kinds of move we invent.

At the International Congress of Mathematicians in Amsterdam in 1954, Wolfgang Haken announced that he could answer the knot triviality problem: it was solvable. However, it took him several years to fill in the details of his argument. Haken's algorithm [85] is based not on diagrams but on the structure of the knot exterior — a compact 3-manifold. Cutting the manifold along a carefully chosen surface simplifies the situation. After repeating the process a finite number of times, all the manifold fragments are balls. The topology of the manifold can be described by listing which surfaces are used for cutting, and how the balls must be glued back together. In the case of a knot exterior, the cutting surface is a minimal-genus surface whose boundary is the given knot. Since it is the only knot spanned by a disc, the trivial knot can be identified.

Haken's method can also be used to attack the knot equivalence problem. The problem is to understand the gluing maps required to reassemble the manifold from the collection of balls. This was achieved in 1979 by Geoffrey Hemion [89], who devised an algorithm to classify the homeomorphisms of a surface. With this information, some 3-manifolds, including all knot exteriors, could now be identified and distinguished. This did not quite solve the knot equivalence problem because it is possible (in theory) for two different knots to have the same exterior. This last snag can be overcome by observing what happens to a meridian of the knot. However, this trick is no longer required: in 1989 Cameron Gordon and John Luecke proved that distinct knots cannot have homeomorphic exteriors [83].

These solutions to the knot problems use 3-manifold techniques, so the problem of detecting triviality using Reidemeister moves remains. For each awkward diagram in Figure 3.23, we know that the complexity must be increased before it can be simplified to a diagram with no crossings. This leads to uncertainty about how bad things could get: is there a limit to the number of crossings that need to be added before we can transform a diagram to a circle? In the late 1990s Joel Hass and Jeffrey Lagarias [87] managed to place a bound on the increase needed in terms of the number of crossings in the starting diagram. Any compact 3-manifold can be built by gluing a finite number of tetrahedra together; this can be done so that a given polygonal knot embedded in the manifold lies in the edge skeleton. Hass and Lagarias showed that an embedding of the trivial knot can be converted into a triangle by applying a bounded number of Δ-moves within this structure. As a consequence, a diagram of the trivial knot can be converted into one with no crossings by applying a bounded number of Reidemeister moves.

Unfortunately, the fact that a problem is solvable does not mean that we have an efficient method for solving it. Haken's algorithm is too complex to implement in practice, and the Hass–Lagarias bound on the number of Reidemeister moves needed for a diagram with a single crossing is larger than the number of atoms in the observable universe. However, it is satisfying to know that we can, in principle, tell knots apart. Two more tractable approaches are discussed on page 283 at the end of the book.

In any case, having an algorithm for detecting knot equivalence produces only a very crude classification — a mere list with no underlying structure. For comparison, observe that the equivalence problem for finite groups can be solved by comparing multiplication tables, but this information does not enable us to answer all the interesting questions in finite group theory. The study of the structural properties of knots and links leads naturally to link invariants. If there are some particular links that we want to distinguish, we can often find a simple invariant to do the job. We shall see examples of this throughout the rest of the book, and show that the list of knots in Appendix A contains no duplicates.

3.10 Exercises

1. Show that the Kinoshita–Terasaka knot (Figure 4.12) has an 11-crossing diagram.

2. Show that the two knots in Figure 3.12 are equivalent.

3. Show that there are finitely many links of each order.

4. Show that an ascending diagram represents a trivial link, and deduce that any diagram can be converted into a diagram of a trivial link by switching crossings.

5. Show that $u(K) \leqslant \frac{1}{2}c(K)$ for any knot K.

6. Let K be a knot with $u(K) = n$. Show there is a diagram D of K such that switching n crossings in D produces a diagram of the trivial knot.

7. Show that the invariant of oriented links $w(L)$ defined on page 62 is zero on all links.

8. Show that linking number is an integer.

9. Let $L \subset \mathbb{R}^3$ be a polygonal link with n triangular components so that $p(L) = 3n$. Show that $c(L) \leqslant 3n(n-1)$.

10. The pretzel knot $P(-1,3,3)$ is a 7-crossing diagram of a much simpler knot. Identify it.

11. Show that, in order for $P(p,q,r)$ to be a knot, either all of p, q and r are odd, or exactly one of them is even. How many components are formed in the other cases?

12. Show that the twist knot $C(-n, -2)$ can be presented as a pretzel knot in two ways: $P(n,1,1)$ and $P(n-1,-3,1)$.

13. Show that the regions of a link projection can be coloured black and white so that no two adjacent regions have the same colour. Such a colouring is called a *chessboard colouring*.

14. Given a chessboard colouring, we can interpret the regions of either colour as a surface whose boundary is the link. The surface looks like a small twisted square near each crossing. The black surface and the white surface can be orientable or unorientable independently. Demonstrate this using the Hopf link, the trefoil, and the figure-8 knot.

15. Show that winding number is an integer.

16. Is there a diagram with writhe zero and winding number zero? Is the answer different for knots and links?

17. For this exercise, you are only allowed to use the Reidemeister moves shown in Figure 3.17.

  (a) Show how to get the mirror image of the type 1 move. (Hint: study Figure 3.22.)

  (b) Show that the mirror image of the type 3 move can be obtained using a sequence of the given type 2 and type 3 moves. One of the intermediate steps may look like this.

18. Fill in the details of the proof of Theorem 3.8.4.

19. Try to find your own awkward diagram of the trivial knot (see Figure 3.23).

4 Constructions and Decompositions of Links

In this chapter we look at some techniques for breaking up links into simpler pieces. These pieces may be complete links, or link fragments known as tangles. When the pieces are also links, the decomposition process can be reversed to give a method for generating new links from given ones.

The primary tool for decomposing links is to embed surfaces in the link exterior. Here we shall explore decompositions using closed surfaces (spheres and tori) and also annuli and other punctured spheres.

4.1 Unions of links

The simplest, if somewhat trivial, decomposition of a link is that of distant union or splittability — the components can be totally disentangled from one another and put on separate planets.

Definition 4.1.1 (split link). A link L is *split* if there is a 2-sphere S embedded in the link complement $\mathbb{R}^3 - L$ so that there are some components of L on each side of S. If we denote the two components of $\mathbb{R}^3 - S$ by U_1 and U_2 and let $L_i = U_i \cap L$ then we write $L = L_1 \sqcup L_2$ and say that L_1 and L_2 are the split components of L.

Conversely, given the two links L_1 and L_2 we can construct L by placing them in disjoint balls in $\mathbb{R}^3$. In this case we say that L is formed as the *distant union* of L_1 and L_2.

We often restrict attention to non-split links.

4.2 Satellites and companionship

Suppose we have a torus embedded in a link complement. If the torus is compressible then either a longitude of the torus bounds a disc in the link complement and the torus is unknotted, or a meridian bounds a disc in the link complement and the link does not travel around the tube of the torus. Neither of these situations is interesting. Another degenerate case occurs when the torus forms the boundary of a solid tubular neighbourhood of one component of the link — this is known as *peripheral tubing*.

It is possible that these are the only ways to embed a torus in the link complement. If there are any other tori then the link can be built from simpler pieces.

Definition 4.2.1 (satellite construction). Let W be a solid torus. A disc properly embedded in W whose boundary is an essential loop in ∂W is called a *meridional disc*. A (possibly knotted) simple loop $\lambda \subset W$ is said to be *essential* if it meets every meridional disc in W. Let $P \subset W$ be a link embedded in an unknotted solid torus in such a way that at least one component of P is an essential loop in W. Let C be a knot and let V denote a solid tubular neighbourhood of C. Choose a homeomorphism $h : W \longrightarrow V$. Then the image $S = h(P)$ is a new link, which is called a *satellite* with *companion* C and *pattern* P. The solid torus V or its boundary surface ∂V is often called the *companion torus* of S.

There is a meridian and longitude coordinate system on the boundary of an embedded solid torus V: a *meridian* is an essential loop in ∂V that bounds a disc in V, and a *longitude* is a simple loop in ∂V that meets a meridian once. Up to isotopy the meridian is unique but there is a choice of longitude. The *framing* of the coordinate system is the linking number

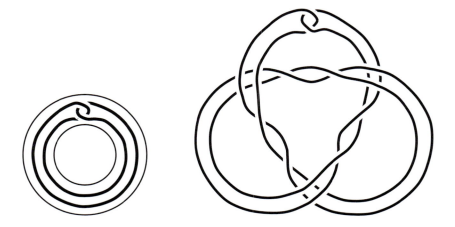

Figure 4.1. Forming a satellite — the untwisted double of a trefoil.

Figure 4.2. A cable of the trefoil.

of the chosen longitude with the core of V. If the framing is zero, the longitude is sometimes called the *preferred* longitude. If h maps a meridian and preferred longitude of W onto a meridian and preferred longitude of V, it is called a *faithful* homeomorphism.

A satellite formed with the pattern shown on the left of Figure 4.1 is called the *double* of C. If h is faithful it is an *untwisted double*. (Note that the pattern is unknotted in $\mathbb{R}^3$.)

If the pattern is a (p, q) torus knot and h is faithful, the satellite is called a (p, q) *cable* of C — see Figure 4.2.

Theorem 4.2.2. *The trivial knot has no non-trivial companions.*

PROOF. Suppose that the trivial knot K has a companion C and associated companion solid torus V, and let $T = \partial V$. Since K is trivial, we can assume that it is embedded in a 2-sphere S.

We now consider the intersection of T and S. Assume that T is isotoped in $\mathbb{R}^3$ so that it meets S transversely. Since T and S are both closed surfaces, the intersection must be a set of simple loops. Since $K \subset S$ and K is an essential loop in V, we have $T \cap S \neq \varnothing$.

Choose a loop $\lambda \in T \cap S$ which bounds an innermost disc Δ in S so that $\Delta \cap T = \partial \Delta = \lambda$. As there are at least two innermost discs and $T \cap K = \varnothing$, it is possible to choose Δ so that $\Delta \cap K = \varnothing$.

If λ bounds a disc in T then we can perform surgery to simplify the situation (see §2.8): cut T along λ and attach a copy of Δ to each boundary. The result of the surgery is a sphere (which we discard) and a torus which we now call T. This procedure reduces the number of intersections of T with S; it can be repeated until there are no loops in $T \cap S$ that bound discs in T.

There are now two cases to consider.

1. Suppose Δ is inside V. Then λ is a meridian of T and Δ is a meridional disc which does not meet K. This is a contradiction since every such disc in T meets K in at least one point.

2. Suppose Δ is outside V so that λ is a longitude of T. Any longitude of T is ambient isotopic in $\mathbb{R}^3$ to the companion, C. Since λ is spanned by the disc Δ, it must be unknotted, and hence, C must be the trivial knot.

$\square$

Definition 4.2.3. A companion of a non-trivial link is a *proper* companion if it is not the trivial knot, and is not equal to any of the link's components.

Links with no proper companions could be called lonely, but, in fact, they are usually called atoroidal — a property of the link complement. The more restrictive term simple, which will be defined in the next section, is more widely used.

4.3 Factorising links

We now try to capture the idea of tying a sequence of knots one after the other.

Let S be a 2-sphere which meets a link L transversely in exactly two points. Then the closure of $S - L$ is an annulus properly embedded in the link exterior $\mathrm{ext}(L)$.

Let $\alpha \subset S$ be an arc in S that connects the points of $L \cap S$. (The choice of arc does not matter since all such arcs are isotopic in S.) Let U_1 and U_2 be the two components of $\mathbb{R}^3 - S$. Define two new links as follows:

$$L_i \;=\; (L \cap U_i) \cup \alpha \quad \text{for } i = 1 \text{ and } 2.$$

We say that S is a *factorising sphere* for L, and that L is a *product* link with *factors* L_1 and L_2. This is written symbolically as $L = L_1 \# L_2$. An example is shown in Figure 4.3.

Figure 4.3. A factorising sphere for the knot $3_1^* \# 4_1$.

Each factor of a product link is also a companion of it: the companion torus is called a *swallow–follow torus* because it swallows one factor and follows the rest of the component as peripheral tubing (see Figure 4.4).

The trivial knot is a factor of every link.

Theorem 4.3.1. *The trivial knot has no non-trivial factors.*

PROOF. The factors of a knot are also its companions. The trivial knot has no non-trivial companions (Theorem 4.2.2) and hence no non-trivial factors.
□

A factor of a link is a *proper* factor if it is not the trivial knot, and is not equal to the link itself. A link with proper factors is called *composite*. A link with no proper factors is called *locally trivial*.

Definition 4.3.2 (prime link). A link is *prime* if it is non-trivial, non-split, and locally trivial.

We often restrict attention to prime links.

Definition 4.3.3 (simple link). A non-trivial link is *simple* if it is prime and has no proper companions.

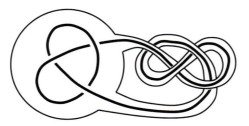

Figure 4.4. A swallow–follow torus for the knot in the previous figure.

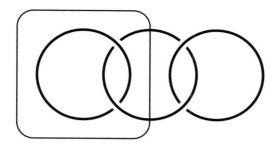

Figure 4.5. A factorising sphere for a 3-component chain.

Any knot with no proper companions is prime (Exercise 4.11.5). For links this is not the case. Figure 4.5 shows a non-prime link which has no proper companions: it is atoroidal but not simple.

The relationships between the various definitions, and others from §4.8, are shown in Figure 4.6.

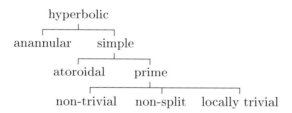

Figure 4.6. Relationships between various link properties.

4.4 Prime satellites

Let us construct some examples of prime links. We shall show that certain satellite links are prime.

Let W be an unknotted solid torus containing a pattern link P. One measure of the complexity of a pattern is the minimum number of intersections P makes with any meridional disc of W. This can be considered either absolutely or algebraically, in which case P is oriented and the sign (or direction) of the intersection is taken into account. The minimum absolute intersection number is called the *wrapping number* of the pattern. The algebraic intersection number is called the *winding number* of the pattern (in this case we do not need to take a minimum since the algebraic intersection number is the same for all meridional discs in general position with respect to P). For example, the pattern in Figure 4.1 has wrapping number 2 and winding number 0.

If the wrapping number is 0 then the pattern sits inside a ball in W and nothing happens when we form the satellite: $S = P$. If the wrapping number is 1 then the companion torus is a swallow–follow torus and the satellite construction produces a product link. If the wrapping number is at least 2 then we say that $P \subset W$ is a *proper pattern*. If a satellite has a proper pattern and the companion knot is non-trivial then S is a *proper satellite*.

Theorem 4.4.1. *A proper satellite is prime if its pattern is a prime knot or the trivial knot.*

PROOF. The proof is similar in style to that of Theorem 4.2.2.

Let K be a satellite knot with companion solid torus V and pattern $P \subset W$. Since K is a proper satellite, every meridional disc of V meets K in at least two points. Let S be a factorising sphere which decomposes K as a product. We assume that K, S and ∂V are in general position.

We now consider the intersection of two surfaces: the sphere S and the torus ∂V. Since these are both closed surfaces, the intersection must be a (possibly empty) set of loops.

Suppose that $S \cap \partial V \neq \varnothing$, and choose a loop $\lambda \in S \cap \partial V$ which bounds an innermost disc Δ in S such that $\Delta \cap \partial V = \partial \Delta = \lambda$. Since there are at least two innermost discs, and $K \cap S$ contains only two points, it is possible to choose Δ so that it meets K in at most one point.

If λ bounds a disc in ∂V then we can perform surgery to remove the intersection, as was done in the proof of Theorem 4.2.2. So we can assume that no loops in $S \cap \partial V$ bound discs in ∂V.

There are now two cases to consider.

1. Suppose Δ is outside V and λ is a longitude of ∂V. Any longitude of ∂V is isotopic to the companion knot which, since it is spanned by the disc Δ, must be the trivial knot. This contradicts the fact that K is a proper satellite.

2. Suppose Δ is inside V and λ is a meridian of ∂V. Since every meridional disc of V meets K in at least two points, Δ must contain both points of $K \cap S$, contrary to assumption.

Both cases lead to a contradiction so there is no such loop λ, and $S \cap \partial V = \varnothing$.

Therefore, S lies inside V, and its preimage $h^{-1}(S)$ in W decomposes the pattern P as a product. This is also a contradiction, and the factorising sphere cannot exist. □

Corollary 4.4.2. *Doubles of non-trivial knots are prime.*

PROOF. The pattern is the trivial knot (see Figure 4.1). □

Corollary 4.4.3. *Cable knots are prime.*

PROOF. Torus knots are prime (Exercise 4.11.10). □

4.5 Uniqueness of factorisation

The factorisation of links into prime links has analogies with the factorisation of numbers. Any positive integer can be written as a product of a finite number of prime numbers and, for the given integer, the set of primes is unique. In the same way a link can be factorised into a finite number of prime links, and these prime factors are uniquely determined.

To simplify things, we shall concentrate on knots: the results, if not these proofs, generalise to links.

First we show that knot factorisation is finite.

Theorem 4.5.1. *A knot has a finite number of factors.*

PROOF. We need to define a complexity function χ from the set of knots to a discrete ordered set (like $\mathbb{N}_0$) with the following properties:

1. $\chi(\bigcirc) = 0$,
2. $\chi(\bigcirc) < \chi(K)$ whenever K is not the trivial knot,
3. $\chi(K_1) < \chi(K_1 \,\#\, K_2)$ whenever K_2 is not the trivial knot.

This means that the complexity is reduced each time we factorise a knot. The complexity is bounded below and is discrete-valued so the factorisation must terminate after a finite number of steps.

We shall construct such a function in Chapter 5: Theorem 5.6.1 shows that the link invariant known as genus has the required properties. □

Note that property 1 is just a normalisation. If we have a complexity function $\widehat{\chi}$ which satisfies properties 2 and 3 we can obtain one whose value is zero on the trivial knot by defining $\chi(K) = \widehat{\chi}(K) - \widehat{\chi}(\bigcirc)$.

We can also deduce Theorem 4.3.1 from such a complexity function: if $K_1 \# K_2 = \bigcirc$ then $\chi(\bigcirc) < \chi(K_1) < \chi(K_1 \# K_2) = \chi(\bigcirc)$, which is a contradiction.

Now we shall show the knot theory equivalent of the statement 'If $a|bc$ then $a|b$ or $a|c$'. That is, if something divides a product, it must divide one of the factors.

Theorem 4.5.2. *Let K be a knot which factorises as $K_A \# K_B$. Let K_P be a prime knot which is a factor of K so that K can also be decomposed as $K_P \# K_Q$. Then one of the following holds.*

1. *K_P is a factor of K_A and K_B is a factor of K_Q: $K_A = K_P \# K_C$ and $K_C \# K_B = K_Q$ for some knot K_C,*

2. *K_P is a factor of K_B and K_A is a factor of K_Q: $K_B = K_P \# K_C$ and $K_C \# K_A = K_Q$ for some knot K_C.*

PROOF. Let U be a ball such that ∂U is a sphere which factorises K as $K_P \# K_Q$, and $K \cap U$ becomes the factor K_P when completed by an arc in ∂U. Let S be the sphere that factorises K as $K_A \# K_B$.

We consider the intersections of the two spheres S and ∂U. If $S \cap \partial U = \varnothing$, the result holds and there is nothing to prove.

So we assume that S and ∂U are in general position. The two spheres intersect in a set of loops, disjoint from each other, and also from the knot: $K \cap S \cap \partial U = \varnothing$. This means that the four points $(K \cap S) \cup (K \cap \partial U)$ are distinct.

Suppose $\lambda \in S \cap \partial U$ is a loop that is innermost on S, and bounds a disc $\Delta \subset S$ such that $\Delta \cap \partial U = \partial \Delta = \lambda$ and $\Delta \cap K = \varnothing$. The loop λ bounds two discs in ∂U, one of which must contain both points of $K \cap \partial U$. Let Δ_U be the other disc so that $\Delta_U \subset \partial U$, $\partial \Delta_U = \partial \Delta$ and $\Delta_U \cap K = \varnothing$. Now $\Delta_U \cup \Delta$ is a sphere that does not meet K. The disc Δ_U can be isotoped in $S^3 - K$ by pushing it across the ball bounded by $\Delta_U \cup \Delta$ until it lies just beyond Δ. This removes the loop λ from the set $S \cap \partial U$.

Hence we can assume that all loops in $S \cap \partial U$ have linking number ± 1 with K (Exercise 4.11.8). This means that, on each factorising sphere, the loops are arranged in parallel and separate the two points of K, rather like lines of latitude separating the two poles on a globe. Thus S is divided into two 'polar' discs and some annuli.

Suppose one of the discs lies inside U — that is, there is a disc $\Delta \subset S$ such that $\Delta \subset U$ and $\Delta \cap \partial U = \partial \Delta$. The disc Δ meets K in a single point, and it divides U into two arc–ball pairs. Since K_P is prime, one of these arc–ball pairs must be trivial and Δ can be isotoped to lie outside U without affecting the decompositions of K into $K_A \# K_B$ and $K_P \# K_Q$.

Thus we may assume that every component of $S \cap U$ is an annulus. Let $R \subset (S \cap U)$ be an annulus which is outermost in U, meaning that it is furthest from K. The two loops of ∂R also bound an annulus $R_U \subset \partial U$. Together, the two annuli form a torus $T = R \cup R_U$. If this torus bounds a solid torus in U then R is parallel to R_U and can be isotoped in $S^3 - K$ to lie outside U, thus reducing the number of intersections of S with ∂U. Otherwise T bounds a 'ball with a knotted hole' in U and is a swallow–follow torus for K which follows the arc $K \cap U$ and swallows the rest of K. That is, T follows K_P and swallows K_Q.

Let U_A be the ball bounded by S such that $K \cap U_A$ becomes the factor K_A when completed by an arc in S. By a small isotopy in $S^3 - K$, R can be moved slightly so that $R \cup R_U$ does not meet S. Thus the swallow–follow torus lies entirely on one side of S. Suppose that $T \subset U_A$. Then T is also a swallow–follow torus for K_A and the followed factor, K_P, must be a factor of K_A. Let K_C denote the factor of K_A that is swallowed by T so that $K_A = K_P \# K_C$. In S^3, the torus swallows K_C and K_B. Hence $K_Q = K_C \# K_B$. □

With the natural numbers, once we have found a common factor we can divide by it to reduce the scale of the problem. With knots, there is no concept of division. In its place, we use the following cancellation rule.

Theorem 4.5.3. *Let K_P be a prime knot and suppose $K_P \# K_Q = K_A \# K_B$. If $K_P = K_A$ then $K_Q = K_B$.*

PROOF. Applying Theorem 4.5.2 we have to consider two cases. First suppose there is a knot K_C such that $K_A = K_P \# K_C$ and $K_C \# K_B = K_Q$. Using the complexity function from the proof of Theorem 4.5.1 on the factorisation of K_A we get

$$\chi(K_P) = \chi(K_A) = \chi(K_P \# K_C) \geqslant \chi(K_P)$$

with equality only when K_C is the trivial knot. Hence $K_Q = \bigcirc \mathbin{\#} K_B = K_B$.

For the second case, we suppose there is a knot K_C such that $K_B = K_P \mathbin{\#} K_C$ and $K_C \mathbin{\#} K_A = K_Q$. Since we are assuming that $K_A = K_P$ we see that K_B and K_Q have the same factors: both are equal to $K_P \mathbin{\#} K_C$. Hence[*] $K_B = K_Q$.   □

From the two preceding theorems we can deduce the following 'uniqueness of factorisation' theorem.

Theorem 4.5.4.   *The factors of a knot are uniquely determined up to order.*

PROOF. Suppose that a knot K is factorised into prime knots in two different ways:

$$K \;=\; A_1 \mathbin{\#} A_2 \mathbin{\#} \cdots \mathbin{\#} A_m \qquad \text{and} \qquad K \;=\; B_1 \mathbin{\#} B_2 \mathbin{\#} \cdots \mathbin{\#} B_n.$$

We need to show that $n = m$ and that there is a permutation π of the integers $1, \ldots, n$ such that $A_i = B_{\pi(i)}$.

If K is the trivial knot then (by Theorem 4.3.1) $m = n = 0$.

Suppose that $m = 1$. By Theorem 4.5.2, A_1 must be a factor of B_1 or of $B_2 \mathbin{\#} \cdots \mathbin{\#} B_n$. Repeating this argument inductively we see that A_1 must be a factor of B_i for some i and, since B_i is prime, we actually have $A_1 = B_i$. Now, by Theorem 4.5.3, we can cancel A_1 and B_i from the equation. The remaining parts must have the same factors, which gives us

$$\bigcirc \;=\; B_1 \mathbin{\#} \cdots \mathbin{\#} B_{i-1} \mathbin{\#} B_{i+1} \mathbin{\#} \cdots \mathbin{\#} B_n.$$

By Theorem 4.3.1, all the B_j with $j \neq i$ must be trivial and, in fact, $m = n = 1$ and $A_1 = B_1$.

We can repeat this argument for the general case and use induction on m to complete the proof.   □

4.6   The product operation

Factorisation has a partial inverse. Suppose that we want to form a link L with given factors L_1 and L_2. Take the split link $L_1 \sqcup L_2$ and let S be

---

[*]Here we must take '=' to mean 'orientation-preserving ambient isotopy'. If it is read as the weaker equivalence relation so that possibly $K_A = -K_P$ then K_B and K_Q need not be the same.

Figure 4.7. Forming a product from 6_3 and 7_6.

a sphere that separates the two constituents. Choose a rectangular disc R whose boundary is composed of four arcs, $\partial R = \{a, b, c, d\}$, such that $L_1 \cap R = a$ and $L_2 \cap R = c$. In addition, we require that $R \cap S$ is a single simple arc, which implies that b and d each meet S in a single point. The product L is formed by switching the arcs in ∂R:

$$L = L_1 \# L_2 = (L_1 - a) \cup (L_2 - c) \cup b \cup d.$$

This is a controlled method of cutting each link and reconnecting the ends.

An example of this product construction is shown in Figure 4.7. The operation (often called *connected sum* because of its relationship to the connected sum of manifolds) is not well-defined in general:

- If both the factors are oriented knots and we form the oriented product then the only choice to be made is where to attach R to the knots. The result is independent of this choice so the product is uniquely defined.

- If orientations on the knots are ignored then there are two ways that R can be attached. If either factor is reversible then both choices lead to the same product. However, if both are non-reversible knots then the two products are distinct.

- Forming the product of links presents yet another choice: which components do we choose to connect?

By most measures of complexity, the product operation tends to make links more complicated. For example, two knots cannot annihilate each other to leave the trivial knot. As a corollary to Theorem 4.3.1 we get

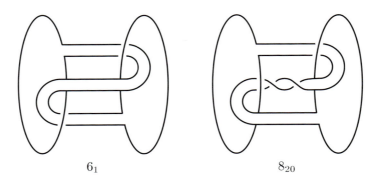

6_1 8_{20}

Figure 4.8. A band-sum of trivial knots can give a non-trivial product.

Theorem 4.6.1. *Given a non-trivial knot K there is no 'anti-knot' K^{-1} such that the product $K \# K^{-1}$ is the trivial knot.*

The set $\mathbb{K}$ of oriented knots under the knot product operation $\#$ can be considered as an algebraic object.

Theorem 4.6.2. *$(\mathbb{K}, \#)$ is an abelian semigroup with unit, and unique factorisation.*

Note that the condition $|R \cap S| = 1$ in the definition of the product operation cannot be removed. Each of the knots in Figure 4.8 is formed by connecting two trivial knots by a rectangular band. However, the 'products' are the prime knots 6_1 and 8_{20}.

4.7   Link invariants and factorisation

When a link can be factorised we can try to find a relationship between invariants of the product and those of its factors. For example, it is easy to see that $\mu(L_1 \# L_2) = \mu(L_1) + \mu(L_2) - 1$. In this case, an explicit formula expresses the number of components of the product in terms of those of the two factors. For other invariants, it may only be possible to find inequalities giving upper or lower bounds.

In the case of unknotting number, it is clear that $u(K_1 \# K_2) \leqslant u(K_1) + u(K_2)$ since each factor can be transformed independently, and the product of two trivial knots is unknotted. Constructions such as this are often possible and give an upper bound. The problem is to show that this produces the best possible solution. For unknotting number, it is conjectured that this number of transformations cannot be reduced and that the formula should

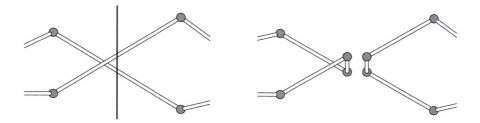

Figure 4.9. Piecewise linear forms of a product and its two factors.

be an equality. There is some evidence to support this. It is known that knots with unknotting number 1 are prime [105] — a fact which could be derived from the conjecture if it were true.

The other link invariant that was introduced in Chapter 1 is the polygon index. In this case, we can also obtain an upper bound by construction, though it is less obvious how to proceed. Let us look at what happens in a simple case.

Suppose that a polygonal knot can be factorised, and that the two factors can be placed on either side of a plane that meets the knot in two points. The sketch on the left of Figure 4.9 illustrates this situation: the plane runs vertically and intersects two edges as shown. The other edges of the link are not included in the sketch since they are not of interest. The sketch on the right shows how polygonal representations of the two factors can be constructed; the edges outside the sketch remain unchanged, and the two edges that meet the plane are cut and reconnected with two additional edges.

We want to be able to reverse this process, starting with a minimal polygon for each factor. In this example, going from the separated state on the right to the product on the left reduces the number of edges by four: one edge is deleted from each factor, and two pairs of edges are merged. As the polygonal representations of the two factors are minimal, we have a good candidate for a minimal polygon of their product. For this construction to be possible, the configuration of edges must have the following properties.

1. The deleted edges must be the same length.

2. The two pairs of merged edges must be suitably aligned.

3. The two factors must be kept separate so that their other edges do not become entangled. This is assured here because the deleted edges have been chosen to lie in the outskirts of their respective factors.

We shall now make these ideas more precise with a few definitions. First we shall deal with property 2: alignment.

Let e_1, e_2 and e_3 be a sequence of consecutive edges in a polygon. If these three edges are not coplanar, they determine a sense of 'twist' which can be associated with the edge e_2. Rotate the polygon so that your line of sight is aligned with the edge e_2. This edge will be seen end on (and hence become invisible) so that the edges e_1 and e_3 appear to have a common vertex. The smallest angle between these edges measured from e_1 to e_3 turns either clockwise or counter-clockwise. We say that edge e_2 has a clockwise or counter-clockwise twist accordingly. If we reverse the orientation of the polygon, so that we are now looking down e_2 from the opposite end, the sense of twist is unchanged. If e_1, e_2 and e_3 are coplanar, we can move one of the vertices at the end of e_2 to produce whichever twist we choose.

Lemma 4.7.1. *Let K be a non-trivial polygonal knot in $\mathbb{R}^3$, and let e_1, e_2 and e_3 be a sequence of consecutive edges in K. Then there is an isotopy of $\mathbb{R}^3$ which deforms K so that e_2 is perpendicular to both e_1 and e_3, and the projections of e_1 and e_3 are perpendicular when projected along e_2.*

PROOF. Since the knot is non-trivial, it must have more than three edges. If all the edges e_1, e_2 and e_3 are coplanar, move one end of e_2 out of the plane. If the displacement is small enough, the knot type will be preserved.

Let Q_1 be the plane spanned by e_1 and e_2. There is an orientation-preserving linear transformation $\Phi_1 : \mathbb{R}^3 \longrightarrow \mathbb{R}^3$ that is the identity on Q_1 and such that $\Phi_1(e_3)$ is perpendicular to Q_1. This linear map can be produced by an isotopy so it preserves the knot type and the sense of the twist along e_2.

Let Q_3 be the plane spanned by $\Phi_1(e_3)$ and e_2. There is another linear map $\Phi_3 : \mathbb{R}^3 \longrightarrow \mathbb{R}^3$ that is the identity on Q_3 and such that $\Phi_3(e_1)$ is perpendicular to Q_3. Again the knot type and twist are preserved. $\square$

Let x be a vertex or an edge of a non-split polygonal link L. Say x is *external* if there is some plane Q such that $L \cap Q = x$. This means that the whole link lies on one side of Q.

Theorem 4.7.2. *If a non-split link can be factorised as $L_1 \# L_2$, some minimal polygon for L_1 has an external edge $\widehat{e}$, and some edge e_z in a minimal polygon for L_2 has the same twist as $\widehat{e}$, then*

$$p(L_1 \# L_2) \leqslant p(L_1) + p(L_2) - 4.$$

PROOF. First apply Lemma 4.7.1 to the external edge $\widehat{e}$ and the two edges adjacent to it in L_1. There is some positive $r \in \mathbb{R}$ such that a (round) ball of radius r contains L_1.

Let e_x, e_z and e_y be a sequence of adjacent edges in L_2. By applying Lemma 4.7.1 and rotating the result, we can arrange that e_x and e_z coincide with the x- and z-axes respectively, and that e_y is parallel to the y-axis. If any other edge of L_2 meets the z-axis, a vertex can be moved slightly to remove the intersection but retain the link type.

Apply a linear transformation to L_2 which stretches it in the z direction so that e_z is the same length as $\widehat{e}$. Let $D = \{(x,y) : x^2 + y^2 < r^2\}$ be a disc in the xy-plane. We can scale L_2 by a linear transformation that stretches it in the x and y directions so that only e_x, e_y and e_z meet the infinite cylinder $D \times \mathbb{R}$. The directions of e_x and e_y are preserved, and e_z remains fixed.

We can now position L_1 and L_2 so that edges e_z and $\widehat{e}$ coincide and the edges adjacent to them are aligned. Deleting e_z and $\widehat{e}$ produces a polygon for $L_1 \mathbin{\#} L_2$ and reduces the edge count by four. $\qquad\square$

The torus knots constructed from hyperboloids in §1.5 provide examples of polygons with no external edges. This makes it unclear whether this theorem can always be applied: is there always a minimal polygon with an external edge of the correct twist? We can weaken the conclusion a little to derive the following, more general corollary.

Theorem 4.7.3. *If a knot can be factorised as $K_1 \mathbin{\#} K_2$ then*

$$p(K_1 \mathbin{\#} K_2) \;\leqslant\; p(K_1) + p(K_2) - 3.$$

PROOF. A knot must have an external vertex. Cut the knot K_1 at an external vertex and insert a short edge in such a way that the added edge is external, and the knot type is maintained. The added edge can be inserted to have either twist, so Theorem 4.7.2 applies. $\qquad\square$

This method does not generalise to links because a component of a polygonal link need not have an external vertex.

4.8 Hyperbolic links

Before we move on from our study of embeddings of tori and annuli in link exteriors, we shall make a short digression. (The ideas in this section are not used in the rest of the book so it is safe to skip to the next one if you get stuck.)

One of the standard examples of the quotient topology is the construction of a torus as a quotient of the plane: $T = \mathbb{R}^2/\mathbb{Z}^2$. This gives the torus a geometric structure: it inherits the Euclidean metric from the covering space, $\mathbb{R}^2$. Similarly, an annulus has a Euclidean structure. Compact orientable surfaces of higher genera have the hyperbolic plane, $\mathbb{H}^2$, as a covering space and they inherit a hyperbolic geometric structure. The sphere is a special case and has a geometry of its own. Knowledge of these structures is often a powerful and useful tool.

An analogous situation holds in three dimensions. Many compact orientable 3-manifolds can be given a geometric structure that is based on one of the three standard models: spherical S^3, Euclidean $\mathbb{R}^3$, or hyperbolic $\mathbb{H}^3$. It turns out that many link exteriors can be given a hyperbolic structure. An interesting feature of hyperbolic manifolds is that they have a well-defined volume, and this can be used as a link invariant.

Which link exteriors can be given a hyperbolic structure? We might expect tori or annuli embedded in the link exterior to cause obstructions because they have a Euclidean structure. However, to answer the question fully, we need to make some definitions.

Let T be a torus embedded in a link exterior $\text{ext}(L)$. We say that T is *boundary parallel* if there is a continuous map $h : T \times [0,1] \longrightarrow \text{ext}(L)$ such that $h(T \times \{0\}) = T$ and $h(T \times \{1\})$ is a component of $\partial \text{ext}(L)$. This means that T is the peripheral tubing described earlier.

Let A be an annulus properly embedded in a link exterior $\text{ext}(L)$. We say that A is *boundary parallel* if there is a continuous map $h : A \times [0,1] \longrightarrow \text{ext}(L)$ such that $h(A \times \{0\}) = A$ and $h(A \times \{1\})$ is a subset of a component of $\partial \text{ext}(L)$. This means that A is a section of peripheral tubing.

A link is *atoroidal* if any torus embedded in the link exterior is compressible or boundary parallel.

A link is *anannular* if any annulus properly embedded in the link exterior is compressible or boundary parallel.

We are now able to answer the question. William Thurston's Hyperbolisation Theorem applied to link exteriors states:

Theorem 4.8.1. *Let L be a non-split link. Then $\text{ext}(L)$ is a manifold with a hyperbolic structure if and only if L is atoroidal and anannular.*

Moreover, ext(L) has finite volume if and only if L is not the trivial knot or the Hopf link.

A link is called *hyperbolic* if it is non-split, atoroidal and anannular. At this point, we should look for examples of hyperbolic links. However, since many links are hyperbolic, it is easier to ask what the definition excludes: what sorts of links are not hyperbolic?

If a link is not atoroidal, it is a satellite link. So a hyperbolic link cannot be a satellite.

How can a link fail to be anannular? When restricted to the link exterior, a factorising sphere is a properly embedded annulus. If both factors are non-trivial then the annulus is not boundary parallel. Thus a hyperbolic link must be prime. Both boundary curves of a factorising annulus are meridians of the link; the other possibility is that the boundary curves are longitudes running along a component of $\partial \operatorname{ext}(L)$. This second possibility is realised in torus links. For example, if $K \subset T$ is a torus knot embedded in a torus then the closure of $T - K$ is an annulus properly embedded in $\operatorname{ext}(K)$ but it is not boundary parallel. So a hyperbolic link cannot be a torus link.

Torus links are not satellites: they are prime and have no proper companions. Therefore, every non-split link belongs to exactly one of the following categories:

- torus links;

- satellite links, including non-prime links;

- hyperbolic links.

The table in Figure 4.10 shows how the knots of small orders are distributed among these classes. From this evidence, it appears that most knots are hyperbolic but this is just a feature of 'small' knots. For randomly generated polygons in $\mathbb{R}^3$ with n edges the probability of getting a composite knot [103] or a prime satellite [104] goes to 1 as n goes to infinity.

4.9 Tangle decomposition

We can generalise link factorisation in the following way. Let S be a 2-sphere which meets a link L transversely in $2n$ points. Then the closure of $S - L$ is a $2n$-punctured sphere properly embedded in the link exterior. Because we want S to bound a ball on both sides, we shall think of L as embedded in S^3 and let U be a component of $S^3 - S$. The pair $(U, U \cap L)$ is called

Order	Torus	Satellite	Hyperbolic	Total
0			1	1
3	1			1
4			1	1
5	1		1	2
6			3	3
7	1		6	7
8	1		20	21
9	1		48	49
10	1		164	165
11	1		551	552
12			2 176	2 176
13	1	2	9 985	9 988
14	1	2	46 969	46 972
15	2	6	253 285	253 293
16	1	10	1 388 694	1 388 705
Total	12	20	1 701 904	1 701 936

Figure 4.10. The numbers of torus knots, satellite knots, and hyperbolic knots for the lower orders (data from [36]). All the satellites listed have the trefoil as companion knot and wrapping number 2.

an *n-tangle*: it is a ball containing n disjoint properly embedded arcs and a (possibly empty) set of loops. When $n = 1$ the decomposition reduces to the previous situation: S is a factorising sphere. We shall concentrate on the case $n = 2$.

An n-tangle is *trivial* if it is homeomorphic to a cylinder $D^2 \times [0, 1]$ containing n parallel straight lines, each connecting the top to the bottom (there are no loops in a trivial tangle).

Theorem 4.9.1. *The trivial knot cannot be decomposed into two non-trivial 2-tangles.*

PROOF. Suppose that the trivial knot K meets a 2-sphere S transversely in exactly four points. Let F be the disc spanning K. We assume that F is isotoped so that it is transverse to S. The intersection $F \cap S$ consists of two arcs (properly embedded in F) and some loops.

Choose a loop $\lambda \in F \cap S$ which bounds an innermost disc Δ in F so that $\Delta \cap S = \partial \Delta = \lambda$. Let B be the ball bounded by S which contains Δ.

Suppose Δ separates the two arcs of $K \cap B$. Then $B - \Delta$ is composed of two 1-tangles. Both of these 1-tangles are factors of K, but since the trivial knot has no non-trivial factors, the 1-tangles are unknotted. Therefore, $K \cap B$ is a trivial tangle.

If Δ does not separate the two arcs of $K \cap B$ then we can perform a surgery in the usual way to remove the loop λ from $F \cap S$. Hence we can assume that $F \cap S$ does not contain any loops. The two arcs of $F \cap S$ divide F in a natural way into three sections. The two outermost sections lie in the same ball of $S^3 - S$ and form a trivial tangle.   □

Hence, in any decomposition of the trivial knot into 2-tangles, at least one tangle must be trivial. It is easy to find examples to show that the other tangle need not be trivial (Exercise 4.11.13).

In the case of 1-tangles, we have the following converse: if a knot can be decomposed into two trivial 1-tangles then the knot must also be trivial. For 2-tangles, this is no longer true. Figure 4.11 shows that the trefoil provides a counterexample.

Definition 4.9.2 (rational link). A non-trivial link is *rational* if it decomposes into two trivial 2-tangles.

The class of rational links has been extensively studied. It provides infinitely many examples of non-trivial prime links that are simple enough to be well understood. We shall study them in more detail in Chapter 8 when the reason for the name will become clear. In the following section we shall show how they get their other common name: 2-bridge links.

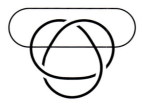

Figure 4.11. The trefoil can be formed from two trivial 2-tangles.

Figure 4.12 shows an example of a knot which is not rational (see the comments on pages 173 and 234, and Theorem 8.7.1). Therefore, whenever it is decomposed into two 2-tangles, at least one of the tangles must be non-trivial; in the decomposition shown, both of the tangles are non-trivial (Exercise 4.11.12).

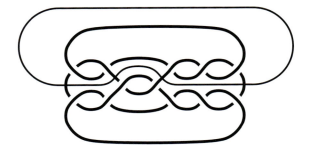

Figure 4.12. The Kinoshita–Terasaka knot (11n42) decomposes into two non-trivial 2-tangles.

4.10 Bridge presentations

Only trivial links can be embedded in a plane, so one way to measure the complexity of a link is to see how 'non-planar' it is. We can position the link so that it lies entirely in a plane except for a few bridges. The bridges must be simple so we insist that their projection onto the plane consists of disjoint straight lines (see Figure 4.13). Such an embedding of a link is called a *bridge presentation*. The minimum number of bridges required is a link invariant called the *bridge number* (or bridge index) of the link and is denoted by $b(L)$.

Bridge number was introduced by Horst Schubert to study companionship. It can be used, for example, as a complexity function to show that every knot has a finite number of companions: if C is a companion of S then $b(C) \leqslant n\, b(S)$ where n is the wrapping number of the pattern [108, 109]. Schubert also established the behaviour of bridge number under factorisation (the case when $n = 1$):

Theorem 4.10.1. *If a knot can be factorised as $K_1 \# K_2$ then*

$$b(K_1 \# K_2) = b(K_1) + b(K_2) - 1.$$

To show that rational links have 2-bridge presentations, we need the following lemma.

Lemma 4.10.2. *Let (U, A) be a trivial n-tangle. There is an isotopy of U keeping ∂A fixed which carries the arcs of A to arcs in ∂U.*

PROOF. Let $D = \{x \in \mathbb{R}^2 : |x| \leqslant 1\}$ be the unit disc and let $I = [0, 1]$ be the unit interval. Let p_i with $1 \leqslant i \leqslant n$ be points in the interior of D, equally

Figure 4.13. Two views of a 2-bridge presentation of the trefoil.

spaced around a circle which is concentric with ∂D. Connect each point p_i with a straight line β_i to its nearest point in ∂D.

The cylinder $D \times I$, together with the n arcs $\alpha_i = p_i \times I$, is a trivial n-tangle. The rectangle $\Delta_i = \beta_i \times I$ is bounded on one side by α_i and on the other three sides by lines in the boundary of the cylinder. These rectangles are disjoint, so there is an isotopy of the cylinder which deforms the arcs without moving their endpoints so that they finish as arcs lying in the boundary of the cylinder: simply slide each arc across its rectangle.

In this picture of a trivial tangle, the deformation is obvious. A general trivial tangle, such as (U, A) can appear to be very messy. However, there is always a homeomorphism of the tangle onto this cylinder (by definition). The inverse images of the discs Δ_i are disjoint in U, and the arcs in A can be deformed in the same way. □

Theorem 4.10.3. *A rational link has a 2-bridge presentation.*

PROOF. Let L be a rational link. There is a sphere S which meets L in four points and defines the two trivial 2-tangles that compose L. We can take S to be the xy-plane plus a point at infinity. Furthermore, we can isotop L so that the four points of $L \cap S$ lie on the x-axis and the two arcs in the upper half-space are semicircles that project down onto the x-axis.

The upper tangle is now arranged in a controlled manner. The lower half-space is a trivial tangle so we can apply the previous lemma: there is an isotopy of the lower half-space which deforms the arcs so that they lie in the xy-plane. The arcs will wind around each other, as can be seen in Figure 4.13. The link now lies entirely in the plane, except for the two semicircular bridges. □

This discussion can be generalised to n-tangles. If a non-trivial link can be decomposed into two trivial n-tangles then the link has an n-bridge presentation, and conversely.

4.11 Exercises

1. Which of the following classes contain a member with reflection symmetry? (Give examples for those that do, and prove impossibility for the others.)

(a) Non-trivial knots,
(b) prime knots,
(c) non-trivial split links,
(d) non-split composite links,
(e) non-split prime links.

2. Show that companionship is reflexive and transitive.

3. Suppose that knots K_1 and K_2 are companions of each other. Show that they are equal up to orientation: $K_1 = K_2$ or $K_1 = -K_2$.

4. Show that the product operation on oriented knots is commutative and associative.

5. Show that a knot with no proper companions is prime.

6. Give an example of a non-prime pattern $P \subset W$ which results in a prime satellite.

7. Assume that the unknotting number conjecture is true: $u(K_1 \# K_2) = u(K_1) + u(K_2)$. Derive the fact that knots with unknotting number 1 are prime.

8. Let S be a factorising sphere for a knot K, and let $\lambda \subset S$ be a simple loop that is disjoint from K. Show that the linking number $\mathrm{lk}(\lambda, K)$ is zero or ± 1.

9. Let L be a non-trivial link such that the removal of any one component produces a trivial link. Such links are called *Brunnian* after Hermann Brunn, who studied them in the early days of knot theory [101].

(a) Show that the Borromean rings are Brunnian.
(b) Find a Brunnian link with n components for any n.
(c) Show that Brunnian links are prime.

10. Show that torus knots are prime. (Hint: look at the intersection of a factorising sphere and the torus holding the knot.)

11. Give an example to show that a decomposition of a prime knot into two non-trivial 2-tangles need not be unique.

12. Given the fact that the trefoil knot is non-trivial, show that both of the tangles in Figure 4.12 are non-trivial.

13. Find a decomposition of the trivial knot into two 2-tangles, one of which is non-trivial.

14. Generalise Theorem 4.9.1 to n-tangles.

15. Show that rational links have one or two components.

16. Show that 1-bridge links are trivial.

17. Show that 2-bridge links are rational and, more generally, an n-bridge link can be decomposed into two trivial n-tangles.

18. Let D be a knot diagram. It can be partitioned into bridges and underpasses so that, at each crossing, the over-crossing strand belongs to a bridge and the under-crossing strand belongs to an underpass. The bridges and underpasses alternate along the knot. Let n be the minimum number of bridges for the given diagram. Show the following [102].

(a) If a bridge crosses an underpass more than once then D does not have minimal crossing number.
(b) If a bridge crosses an adjacent underpass then D does not have minimal crossing number.
(c) If D is a minimal-crossing diagram then $c(D) \leqslant n\,(n-2)$.

19. Show that rational knots are prime. (Hint: look at the intersection of a factorising sphere and the sphere defining the trivial tangles.)

5 Spanning Surfaces and Genus

In this chapter we shall see that the topology of knots and links is closely related to properties of surfaces. A surface F *spans* a link L if its boundary ∂F is ambient isotopic to L.

Definition 5.0.1 (link genus). The *genus* of an oriented link L is the minimum genus of any connected orientable surface that spans L. The genus of an unoriented link is the minimum taken over all possible choices of orientation. We denote the genus of L by $g(L)$.

There are several observations that arise from this definition. First, it is not obvious that genus is well-defined: does every link bound an orientable surface? In the exercises at the end of Chapter 3 you constructed spanning surfaces from chessboard colourings, but these are not necessarily orientable.

Second, genus is yet another invariant defined by minimising a geometric property over all possibilities and therefore it is difficult to calculate in general. Any knot with genus zero is spanned by a disc and hence is the trivial knot. A genus-1 surface spanning a non-trivial knot must be minimal. Proving that any other surface is minimal can be very hard. In Chapter 7 we shall see that link genus can be related to a polynomial invariant which gives a lower bound. We can get an upper bound by construction, therefore some classes of surface can be proved to be minimal.

5.1 Seifert's algorithm

First we establish that link genus is well-defined [116, 122].

Theorem 5.1.1 (Seifert's algorithm). *Every link bounds an orientable surface.*

Figure 5.1. Adding a twisted rectangle in the neighbourhood of a crossing.

PROOF. If the link is unoriented then choose an orientation for each component. Choose a diagram for the link in the xy-plane in $\mathbb{R}^3$. In a small neighbourhood of each crossing, make the following local change to the diagram: delete the crossing and reconnect the loose ends in the only way compatible with the orientation. An example is shown in the first two pictures of Figure 5.1; the broken circle indicates the neighbourhood of activity. When this has been done at every crossing, the diagram becomes a set of disjoint simple loops in the plane — it is a diagram with no crossings. These loops are called *Seifert circles*.

The Seifert circles may be nested. We can assign an index to each one: let $h(\lambda)$ be the number of Seifert circles that contain λ. We intend to use this index as a height function.

We shall now construct a surface. For each Seifert circle λ, take a disc Δ in the plane $z = h(\lambda)$ such that $\partial\Delta$ projects down onto λ. This collection of discs lives in the upper half-space $\mathbb{R}^3_+$ and the discs are stacked in such a way that when viewed from above, the boundary of each disc is visible. The discs inherit an orientation from the diagram. We can colour the two sides of each disc red and black in the standard way: if the boundary of a disc is oriented clockwise when viewed from above then the upper surface is coloured black, otherwise it is red.

To complete the required surface, we need to insert a small twisted rectangle at the site of each crossing, the sense of the half-twist being chosen to produce the correct kind of crossing (see the right of Figure 5.1). This produces a surface with boundary L.

It remains to show that the surface is orientable. If a band connects two discs with the same height index then they must be coloured differently on their upper surfaces. Thus the colour can be extended naturally across the half-twisted band without causing a problem. Any other band will connect two discs whose height indices differ by 1. These discs will have the same colour uppermost, and when the band is added, the half-twist brings the upper side of the lower disc round to match with the upper disc. Therefore, the whole surface is two-sided.

If the diagram is disconnected, the surface will also be disconnected. If a connected surface is required, the components can be piped together so that the orientation is preserved.   □

An orientable surface which spans a link is often called a *Seifert surface*. In this book we shall distinguish between a surface that is constructed by applying Seifert's algorithm, which is called a *projection surface*, and surfaces in general, which are referred to as *spanning surfaces*.

Theorem 5.1.2. *The Euler characteristic of a projection surface F constructed from a diagram D with $s(D)$ Seifert circles and $c(D)$ crossings is*

$$\chi(F) \; = \; s(D) - c(D).$$

PROOF. The projection surface is built from discs and bands. We divide these elements into triangles as shown in Figure 5.2: a disc at which n bands meet is divided into $2n$ triangles with a vertex in its interior, and each band is divided into two triangles. Let J denote the total number of joins where a band is attached to a disc. Each rectangle is attached at two ends so $J = 2\,c(D)$. There are $2J$ triangles in the discs and two in each band so the number of faces in the triangulation is $2J + 2\,c(D)$. Simple counting shows that there are $2J + s(D)$ vertices and $4J + 3\,c(D)$ edges. Hence

$$\chi(F) \; = \; [2J + s(D)] - [4J + 3\,c(D)] + [2J + 2\,c(D)] \; = \; s(D) - c(D).$$

□

Corollary 5.1.3. *The genus of a projection surface F constructed from a connected diagram D satisfies*

$$2\,g(F) \; = \; [1 - s(D) + c(D)] + [1 - \mu(D)].$$

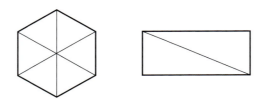

Figure 5.2. Triangulations of a disc where three bands meet, and a single band.

PROOF. Let $\widetilde{F}$ denote the closed surface associated to F. Then

$$
\begin{aligned}
2\,g(F) &= 2\,g(\widetilde{F}) \\
&= 2 - \chi(\widetilde{F}) \\
&= 2 - (\chi(F) + |\partial F|) \\
&= [1 - \chi(F)] + [1 - \mu(D)].
\end{aligned}
$$

$\square$

Note that it is not always possible to construct a minimal-genus surface from a minimal-crossing diagram. For example, the knots 10_{163} and 10_{165} both have genus 2 and projection surfaces with this genus can be constructed from 11-crossing diagrams; the projection surfaces constructed from 10-crossing diagrams all have genus at least 3 [123]. We shall see in Corollary 10.2.6 that for some knots it is impossible to obtain a minimal-genus spanning surface using Seifert's algorithm on any diagram.

5.2 Seifert graphs

We can construct a graph G from a projection surface as follows: associate a vertex with each Seifert circle and connect two vertices with an edge if their Seifert circles are connected by a twisted band. This graph is called a *Seifert graph* [140]. It can be embedded in the projection surface (each vertex lies in a disc, each edge runs through a band), and it can also be embedded in the plane. Examples are shown in Figures 5.3 and 7.4.

Much of the information about a projection surface is encoded in a Seifert graph. Each edge of the graph can be labelled $+$ or $-$ depending on the

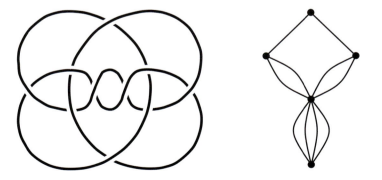

Figure 5.3. The knot 10_{165} and its Seifert graph of two blocks.

sign of its associated crossing in the underlying diagram D. However, this still does not provide enough information to reconstruct the diagram from the graph.

Some of the structural properties of D and G are related. For example, every circuit in G must have even length because the algorithm constructs an orientable surface. This means that G is a bipartite graph: its vertices can be divided into two sets so that each edge has an endpoint in each set. If G contains an isthmus then D is a reducible diagram. This does not force L to be a non-prime link; it only means that D does not have minimal crossing number. Seifert circles are sometimes classified into those that have Seifert circles on only one side (type I) and those that have Seifert circles both inside and outside (type II). A type II Seifert circle gives rise to a cut vertex in G.

Theorem 5.2.1. *Let G be a Seifert graph constructed from a projection surface F and diagram D. Then*

$$\operatorname{rank}(G) \;=\; c(D) - s(D) + 1 \;=\; 1 - \chi(F).$$

PROOF. Suppose G is a graph with V vertices and E edges. A spanning tree in G contains $V - 1$ edges so the rank of G is $E - V + 1$. Substituting $V = s(D)$ and $E = c(D)$ gives the first equality. The equality on the right is just Theorem 5.1.2. □

5.3   Genus-1 knots

Genus is an *intrinsic* property of a surface: it depends on the abstract surface (its homeomorphism class) but not on how the surface is embedded. To give an idea of the variety of possible embeddings we shall consider surfaces of genus 1.

A genus-1 surface spanning a knot is homeomorphic to a standardly embedded genus-1 surface with one boundary component — that is, a torus with a disc removed. The sequence of deformations shown in Figure 5.4 shows that this is isotopic to two annuli stuck together. This surface can also be thought of as a disc with two bands attached, and can be redrawn as shown in Figure 5.5.

This is the fundamental surface of our examples but its boundary is unknotted. The embedding of the surface can be changed in many ways to produce a knotted boundary. A simple example is shown on the left of Figure 5.6: one band passes through the tube formed by the other to

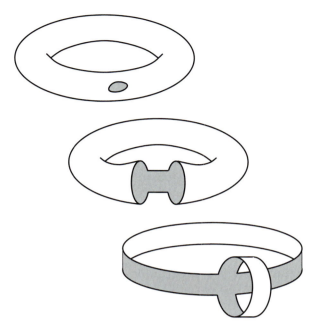

Figure 5.4. Isotopy of a torus minus a disc.

give the knot 6_1. In the example on the right, the band inside the tube is knotted. Here the surface can be further complicated: the tube containing the knotted band can also be knotted so that it follows the band. This gives two different genus-1 surfaces spanning the same knot.

Surfaces in disc–band form (Figure 5.5) can display the features shown in Figure 5.7: the bands can be twisted, knotted, or linked. To keep the same side of the surface uppermost, twists can be replaced by curls. Combining these in different ways gives a great variety of possible embeddings. By contrast, the genus-1 surfaces produced by Seifert's algorithm are of very few types. Diagrams of the knots referred to in the following proposition are shown in Figure 3.8.

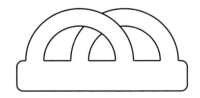

Figure 5.5. A disc–band surface of genus 1.

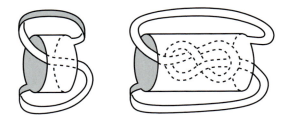

Figure 5.6. Genus-1 surfaces spanning knots.

Theorem 5.3.1. *A knot has a genus-1 projection surface if and only if it is one of the following:*

(a) *the trivial knot,*

(b) *a rational knot $C(2p,\ 2q)$,*

(c) *a (p, q, r) pretzel knot with p, q and r all odd.*

PROOF. Let D be a diagram of a knot with projection surface F and suppose that $g(F) = 1$. A Seifert graph for a genus-1 surface spanning a knot has rank 2. We can simplify such graphs by deleting pairs of adjacent 2-valent vertices. This corresponds to removing full twists from the bands of F. We can repeat this until there are no adjacent 2-valent vertices left. When this happens, every band contains one or two half-twists.

The graphs we can obtain are shown on the left of Figure 5.8. These three graphs lead to four possible arrangements of Seifert circles, but not all the configurations correspond to knots; some are 3-component links and these have genus 0.

Consider the lower graph with three vertices. Each vertex corresponds to a disc in the surface so we expand each vertex to a disc. The edges between them can be inserted in the two ways shown on the right of Figure 5.8. Each edge represents a twist-box containing an odd number of half-twists.

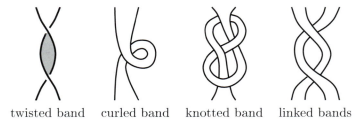

twisted band   curled band   knotted band   linked bands

Figure 5.7. Features that can appear in disc–band surfaces.

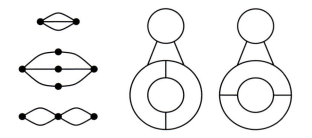

Figure 5.8. Seifert graphs of rank 2 and some circle configurations.

In the first case, the diagram D must have the form of a rational knot $C(2p, 2q)$ — see Figure 3.8(b). The other case gives a product of torus links $T(2, 2p) \# T(2, 2q)$ — see Figure 5.9(a).

Both of the other graphs correspond to pretzel links. The 2-vertex graph gives a (p, q, r) pretzel knot with all the twist-boxes containing an odd number of half-twists — see Figure 3.8(d). For the 5-vertex graph, all the twist-boxes contain an even number of half-twists giving a 3-component pretzel link $P(2p, 2q, 2r)$ — see Figure 5.9(b).

A diagram like $P(1, 1, -1)$ gives a genus-1 surface for the trivial knot.
$\square$

Let K be the untwisted double of a trefoil. It has a spanning surface of genus 1: in the disc–band form one band is knotted in the form of a trefoil and the other band contains a full twist. However, it appears that K does not fit into any of the categories of Theorem 5.3.1, and we shall show that this is indeed the case in Corollary 10.2.6. This means that K does not have a projection surface which is also a minimal-genus spanning surface.

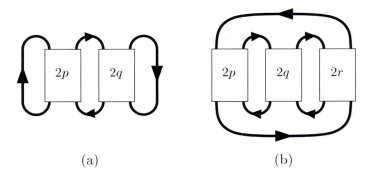

(a)   (b)

Figure 5.9. Some 3-component links that have Seifert graphs of rank 2.

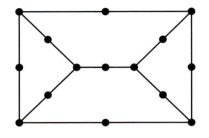

Figure 5.10.

Note that the enumeration method of Theorem 5.3.1 is not suitable for higher genera. Following that route, to find knot diagrams that produce projection surfaces of genus 2, we would first enumerate all Seifert graphs of rank 4 with no adjacent 2-valent vertices, then convert them into diagrams in all possible ways, and finally discard links and non-prime knots.

It is clear that the number of edges in a rank-4 Seifert graph with no adjacent 2-valent vertices is bounded. Indeed, a graph with the upper bound of 18 is shown in Figure 5.10. This is equivalent to bounding the crossing number of the corresponding diagrams. Alexander Stoimenow [123] argued from features of the diagram that, in fact, at most 15 crossings are possible. He then used a computer to search all knot diagrams up to 15 crossings, reduce them by removing full twists, and remove duplicates that are related by flypes. This resulted in the 24 families of diagrams shown in Figure 5.11. The boxes labelled '2' contain an even number of half-twists; the others contain an odd number of half-twists. The knot obtained by substituting the minimum number of half-twists in each box so as to produce an alternating diagram is shown in parentheses beneath each diagram. The knots up to 10 crossings use the classical notation of Rolfsen [5]; the others use the notation of the *Knotscape* database [254]. Only a few of the families have common names: the diagram labelled 5_1 is a generalised pretzel knot, and the diagram labelled 8_{12} is a genus-2 rational knot.

5.4   Different genus measures

Just as an aside, we shall take a quick look at other link invariants that can be derived from spanning surfaces. Placing a variety of restrictions on the spanning surface leads to several distinct notions of link genus.

$g(L)$ The standard genus defined on page 102 is the minimum genus taken over all orientable surfaces spanning L.

$g_c(L)$ The canonical genus is the minimum genus taken over all surfaces constructed by applying Seifert's algorithm to diagrams of L.

$g_b(L)$ The flat genus is a restriction of the canonical genus: it is the minimum genus taken over all surfaces constructed by applying Seifert's algorithm to diagrams of L which do not have nested Seifert circles [117]. This means that the surface is planar except in neighbourhoods of the crossings.

$g_f(L)$ The free genus is the minimum genus taken over all unknotted orientable surfaces spanning L. A surface F is unknotted if the fundamental group $\pi_1(S^3 - F)$ is a free group. All projection surfaces are unknotted; the genus-1 surface spanning a double is not.

$g_4(L)$ The 4-ball genus, or slice genus, is the minimum genus taken over all orientable surfaces spanning L in the 4-dimensional ball. For this construction, we regard the link as a subset of $S^3 = \partial B^4$ so the surface is properly embedded in B^4.

To understand g_4 let us first consider an example in one dimension lower. Suppose two lines in $S^2 = \partial B^3$ intersect transversely in a point. If part of one of the lines in a neighbourhood of the intersection is pushed into the interior of B^3 then the intersection can be removed. Similarly, if two surfaces in $S^3 = \partial B^4$ intersect transversely as shown in Figure 5.12, then the hatched area can be pushed into the 4-ball to remove the intersection.

Surface intersections like the one in Figure 5.12 are called *ribbon singularities*. An immersed surface in S^3 in which all the self-intersections are ribbon singularities is called a ribbon surface. By lifting small regions of a ribbon surface into the 4-ball, it can be made into a non-singular embedded surface in $\mathbb{R}^4$.

Let us construct some examples. It is easy to see that each of the knots in Figure 4.8 can be spanned by two discs connected together by a long, thin, possibly twisted, rectangular strip. In order to do this the strip must pass through each disc once in a ribbon singularity. The ribbon surface (two discs joined by a single strip) is topologically a disc, so each knot bounds a disc properly embedded in B^4 and hence has $g_4 = 0$. Ribbon surfaces of higher genera can be constructed by increasing the number of strips.

Knots that are spanned by a ribbon disc in S^3 are called *ribbon knots*; knots for which $g_4 = 0$ are called *slice knots*. Clearly, every ribbon knot is

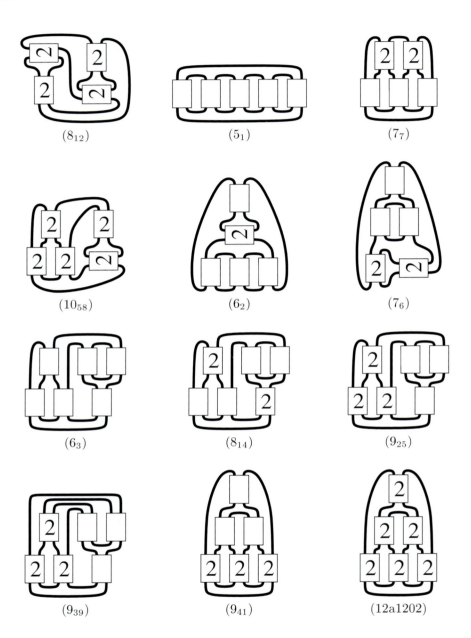

Figure 5.11. The 24 families of knot diagrams with canonical genus 2. Twist-boxes labelled '2' contain an even number of half-twists; the others contain an odd number of half-twists [123].

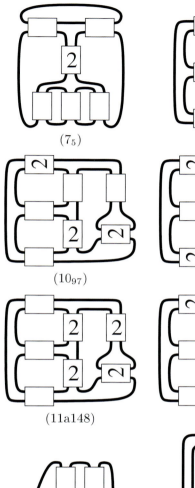

(7_5)

(9_{23})

(9_{38})

(10_{97})

(10_{101})

(10_{120})

$(11a148)$

$(11a123)$

$(11a329)$

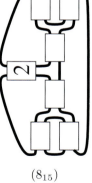

(8_{15})

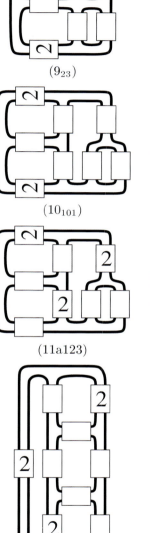

$(12a1097)$

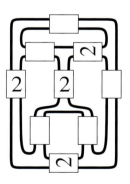

$(13a4233)$

Figure 5.11 (*continued*).

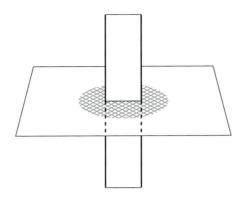

Figure 5.12. A ribbon singularity can be removed by pushing the hatched area into the fourth dimension.

a slice knot. It is conjectured that the converse is also true: that every slice knot is a ribbon knot.

It is easy to see that $g_b(L) \geqslant g_c(L) \geqslant g_f(L) \geqslant g(L) \geqslant g_4(L)$. For some links, such as the figure-8 knot, these genera are all equal. It can be shown that $g_b(L) = g_c(L)$ for all links [117], but the other genus invariants are distinct. The ribbon knots 6_1 and 8_{20} in Figure 4.8 are examples for which $g(L) > g_4(L)$. If K is the untwisted double of the trefoil then $g(K) = 1$ since K is non-trivial, $g_f(K) = 2$ [119], and $g_c(K) = 3$ (see Exercise 5.9.13 and Corollary 10.2.6).

In all these examples we have assumed that the spanning surface is orientable. Yet more variations can be defined by dropping this condition [115, 121].

5.5 Surgery equivalence

Links can have different spanning surfaces. By different we mean that the surface embeddings are not ambient isotopic. If the surfaces have different genera then they are clearly not isotopic. However, it is possible for a link to have distinct spanning surfaces of minimal genus [120].

How are such surfaces related? We know that surfaces can be changed by using surgery to add or remove handles. In fact, we can always use a combination of surgery and isotopy to transform a surface into any other one with the same boundary. An easily accessible combinatorial proof of this result, based on Reidemeister's Theorem (3.8.1), has been produced by Dror Bar-Natan, Jason Fulman and Louis Kauffman [114].

Two surfaces are said to be *S-equivalent* if one can be converted into the other by a finite sequence of tubing and compressing operations.

Theorem 5.5.1. *If F_1 and F_2 are two surfaces such that ∂F_1 and ∂F_2 are equivalent links then F_1 and F_2 are S-equivalent.*

The proof of this theorem will be given in the next four lemmas. First we introduce some notation to reduce the number of figures required. Parallel lines with arrowheads denote short sections of adjacent Seifert circles. Combinations of four symbols will be placed between sets of these lines. The symbols '+' and '−' indicate positive and negative crossings, respectively; '?' indicates a crossing of either type. These can also be thought of as twisted rectangles in the projection surface. A 'T' between the lines indicates that there is a tube between the discs spanning the Seifert circles. So, for example, the diagram fragment shown in Figure 5.13 (ignoring the broken lines) can be summarised as

$$\uparrow\ +\ \uparrow\ \begin{matrix}?\\+\end{matrix}\ \uparrow$$

For two surfaces F_1 and F_2, we shall write $F_1 = F_2$ if they are isotopic, and $F_1 \overset{\text{s}}{\sim} F_2$ if they are S-equivalent.

Lemma 5.5.2. *Surfaces with the following configurations are isotopic (and hence S-equivalent).*

$$\uparrow\ \begin{matrix}+\\-\end{matrix}\ \uparrow\ =\ \uparrow\ T\ \uparrow\ =\ \uparrow\ \begin{matrix}-\\+\end{matrix}\ \uparrow$$

PROOF. Exercise (draw a picture and find the tube). $\qquad\qquad\square$

Lemma 5.5.3. *Surfaces with the following configurations are S-equivalent.*

$$1.\quad \uparrow\ +\ \uparrow\ \begin{matrix}?\\+\end{matrix}\ \uparrow\ =\ \uparrow\ \begin{matrix}+\\?\end{matrix}\ \uparrow\ +\ \uparrow \qquad\qquad 2.\quad \uparrow\ +\ \uparrow\ \begin{matrix}+\\-\end{matrix}\ \uparrow\ \overset{\text{s}}{\sim}\ \uparrow\ \begin{matrix}-\\+\end{matrix}\ \uparrow\ +\ \uparrow$$

$$3.\quad \uparrow\ -\ \uparrow\ \begin{matrix}-\\?\end{matrix}\ \uparrow\ =\ \uparrow\ \begin{matrix}?\\-\end{matrix}\ \uparrow\ -\ \uparrow \qquad\qquad 4.\quad \uparrow\ -\ \uparrow\ \begin{matrix}+\\-\end{matrix}\ \uparrow\ \overset{\text{s}}{\sim}\ \uparrow\ \begin{matrix}-\\+\end{matrix}\ \uparrow\ -\ \uparrow$$

PROOF. The situation corresponding to the left-hand side of case 1 is shown in Figure 5.13. The crossing labelled '?' can be slid along the thickened

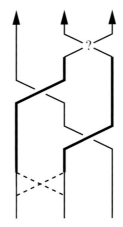

Figure 5.13. The upper crossing slides along.

lines until it reaches the position indicated by the broken lines. This gives the expression on the right-hand side.

Case 2 cannot be achieved by the same kind of surface isotopy (why not?). Instead we proceed as indicated in the sequence below. First add a tube, which is equivalent to adding two bands of opposite signs, then apply case 1, and delete a tube, as follows.

Cases 3 and 4 are similar.   □

A connected orientable surface of genus g with n boundary components is often thought of as a sphere with g tubes (or handles) added and n discs removed. If $n \geqslant 1$ such a surface is also homeomorphic to a disc–band surface like the one shown in Figure 5.14. Along the top of the disc there are g pairs of bands, and along the bottom there are $(n-1)$ single bands. Any

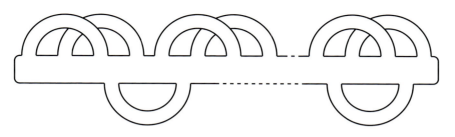

Figure 5.14. A generic disc–band surface.

orientable surface embedded in $\mathbb{R}^3$ can be isotoped to be in disc–band form although, in general, the bands will be twisted, knotted, and interlinked as in Figure 5.7.

Lemma 5.5.4. *A disc–band surface is S-equivalent to a projection surface.*

PROOF. If a band crosses over another band, as shown on the left of Figure 5.15, then the surface can be changed by adding a tube between the two bands. This converts the surface into the corresponding projection surface shown in the centre.

If the surfaces have the same side uppermost at the crossover point, then one of them can be given a half-twist, as shown on the right of the figure; the previous method then applies. □

Lemma 5.5.5. *Any two projection surfaces of the same link are S-equivalent.*

PROOF. To prove this it is sufficient to examine how each of the Reidemeister moves affects the projection surface. In each case, the surfaces before and after a move are either isotopic or S-equivalent.

Figure 5.15. A disc–band surface is S-equivalent to a projection surface.

The situation for type 1 moves is shown in Figure 5.16. The broken circle indicates the boundary of a ball that contains all the changes; everything outside remains fixed. It is clear that the surfaces before and after differ only by a small isotopy — a patch of the surface is just twisted over.

The three configurations to consider for type 2 moves are shown in Figure 5.17 with before and after placed above and below. In the first case, the two arcs belong to the same Seifert circle and the surface is just twisted. In the middle case, the arcs have the same orientation but now bound distinct Seifert circles. The upper picture can be obtained from the lower one by untwisting the band to give a tube, then compressing it. The last case is that of Lemma 5.5.2.

The four configurations to consider for type 3 moves are shown in Figure 5.18, again arranged with before and after placed above and below. In the first three cases, Lemma 5.5.3 can be applied to show S-equivalence. The final case requires an isotopy, application of type 2 moves until Lemma 5.5.3 can be applied, then simplification; the sequence of deformations is shown in Figure 5.19.   □

5.6 Genus and factorisation

We now study how genus behaves under link factorisation.

Theorem 5.6.1. *Suppose that a link L has a connected incompressible spanning surface of minimal genus. If L can be factorised as $L_1 \# L_2$ then*

$$g(L_1 \# L_2) \ = \ g(L_1) + g(L_2).$$

PROOF. Let F_i be a minimal-genus spanning surface for L_i. Take a 2-sphere $S \subset \mathbb{R}^3$ which separates $\mathbb{R}^3$ into two pieces U_1 and U_2 so that $U_1 \cup U_2 = \mathbb{R}^3$ and $U_1 \cap U_2 = \partial U_i = S$, and such that $F_i \subset U_i$ with $F_1 \cap S = F_2 \cap S$ equals a single simple arc. Then $F_1 \cup F_2$ is a spanning surface for $L_1 \# L_2$. Hence,

$$g(L_1) + g(L_2) \ = \ g(F_1) + g(F_2) \ = \ g(F_1 \cup F_2) \ \geqslant \ g(L_1 \# L_2).$$

To show the reverse inequality, we take F to be a connected, incompressible, minimal-genus surface spanning $L = L_1 \# L_2$, and let S be a factorising sphere. Assume that F and S are in general position. The two surfaces intersect in an arc α, connecting the two points of $L \cap S$, and a (possibly empty) set of loops.

Figure 5.16. Type 1 Reidemeister moves give isotopic surfaces.

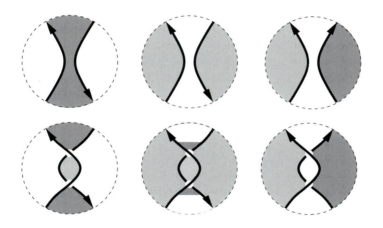

Figure 5.17. The effects of type 2 Reidemeister moves on surfaces.

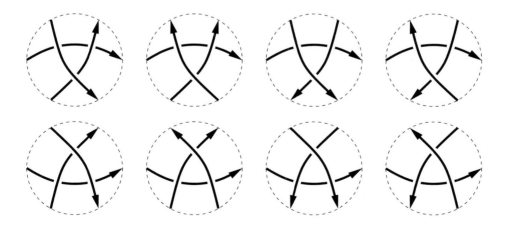

Figure 5.18. The oriented type 3 Reidemeister moves.

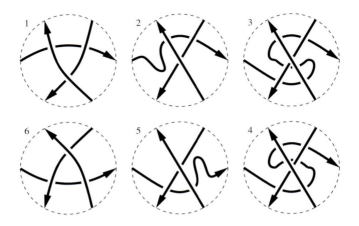

Figure 5.19. S-equivalence: the final case.

Let $\lambda \in F \cap S$ be a loop which is innermost on S, and bounds a disc $\Delta \subset S$ so that $\Delta \cap F = \partial \Delta = \lambda$. There must be such a loop since α lies on only one side of any loop in S.

If λ bounds a disc in F then we can perform surgery to simplify the situation: cut F along λ and attach a copy of Δ to each boundary. The result of the surgery is a sphere (which we discard) and a surface spanning L which we continue to call F. This procedure reduces the number of intersections of F with S; it can be repeated until there are no loops in $F \cap S$ that bound discs in F.

There are two further cases to consider.

1. If λ is separating in F but does not bound a disc in F then Δ is a compressing disc for F.

2. If λ is a non-separating curve in F then, again, Δ is a compressing disc.

Both cases contradict the assumption that F is incompressible, so there is no such loop λ, and $F \cap S = \alpha$.

Let F_1 and F_2 be the two components of $F - \alpha$; they are spanning surfaces for L_1 and L_2. Therefore,

$$g(L_1 \# L_2) \; = \; g(F) \; = \; g(F_1) + g(F_2) \; \geqslant \; g(L_1) + g(L_2).$$

□

Why does the theorem statement contain the incompressibility condition? Without it, to prove the second part we would take F to be a minimal genus surface spanning $L_1 \# L_2$. Then in case 1 surgery of F along λ would produce a disconnected surface. Both components would have a boundary (otherwise one component could be discarded to leave a surface of smaller genus). This would leave us with a link spanned by a disconnected minimal-genus surface. Such links do exist: split links are an obvious example, but there are others (see Exercise 5.9.16).

The following corollary holds because the trivial knot is the unique knot with genus zero. A generalisation to links is not possible: there are composite links of genus zero (see Figure 4.5).

Corollary 5.6.2. *Genus-1 knots are prime.*

Corollary 5.6.3. *Knot factorisation is finite.*

PROOF. See Theorem 4.5.1. □

5.7 Linking number revisited

Given two (possibly knotted) simple oriented loops λ_a and λ_b we can take an orientable surface F_a that spans λ_a. Since F_a is orientable, it has a positive (red) side and a negative (black) side. Assuming that λ_b is in general position with respect to F_a, we can associate a signed intersection number to each point where λ_b punctures F_a: if λ_b passes from the negative side of F_a through to the positive side we assign $+1$ to the intersection; if it passes in the opposite direction we assign it -1. We denote the sum of these intersection numbers by $|F_a \cap \lambda_b|^*$.

It is not immediately clear that this value is independent of the choice of spanning surface, or of which loop it spans. However, the following discussion shows that this is, in fact, the case. Furthermore, it is equal to the linking number $\text{lk}(\lambda_a, \lambda_b)$.

Lemma 5.7.1. *The value $|F_a \cap \lambda_b|^*$ does not depend on the choice of spanning surface F_a.*

PROOF. By Theorem 5.5.1, any two surfaces spanning λ_a are connected by a finite sequence of piping and compressing operations. We shall show that these operations do not change $|F_a \cap \lambda_b|^*$.

Let α be an embedded arc which meets F_a only at its endpoints and let $V = D^2 \times \alpha$ be a solid tubular neighbourhood such that $V \cap F_a = D^2 \times \partial\alpha$. Now $S = \partial V$ is a sphere composed of an annulus $A = \partial D^2 \times \alpha$ and two discs $S - A = D^2 \times \partial\alpha$. For any surgery operation, there is some α such that the surgery corresponds to replacing A by $S - A$ (compressing) or replacing $S - A$ by A (piping).

Note that when we orient the regions of S with the natural orientation they inherit from F_a before and after surgery, the positive sides of A and $S - A$ do not match up: if the annular region A of S has its positive (red) side on the outside of the sphere then the discs of $S - A$ have their negative (black) side on the outside of S.

We now examine how λ_b meets S. The number of times λ_b enters S and the number of times it leaves S must be equal. For notational shorthand we shall write $\#(\lambda$ enters $X)$ to mean 'the number of times the loop λ enters the sphere S passing through the set X'. Similarly, $\#(\lambda$ exits $X)$. Now,

$$\#(\lambda_b \text{ enters } A) + \#(\lambda_b \text{ enters } S - A) \;=\; \#(\lambda_b \text{ exits } A) + \#(\lambda_b \text{ exits } S - A).$$

Rearranging we get

$$\#(\lambda_b \text{ enters } A) - \#(\lambda_b \text{ exits } A) \;=\; \#(\lambda_b \text{ exits } S - A) - \#(\lambda_b \text{ enters } S - A).$$

The left-hand side is the intersection number of λ_b with A. The right-hand side is the intersection number of λ_b with $S - A$: the terms are in the opposite order because the positive and negative sides of A and $S - A$ are opposite. Hence, $|A \cap \lambda_b|^* = |(S - A) \cap \lambda_b|^*$ and performing the surgery does not change the intersection number of λ_b with F_a. $\qquad\square$

The next result is unnecessary as it follows from Lemma 5.7.1, Theorem 5.7.3, and the fact that the linking number is symmetric. However, it shows that linking number can be defined in a 3-dimensional context without the use of diagrams.

Lemma 5.7.2. *The calculation is symmetric: $|F_a \cap \lambda_b|^* = |\lambda_a \cap F_b|^*$. Its value does not depend on the choice of component spanned by the surface.*

PROOF. For this proof, we shall work explicitly in the piecewise linear category. This means that a surface is thought of as built up from a finite number of small flat triangles glued together along their edges.

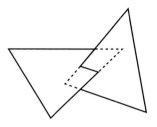

Figure 5.20. Two linked triangles: $|\Delta_i \cap \partial\Delta_j|^* = |\partial\Delta_i \cap \Delta_j|^*$.

Suppose that we have two surfaces: F_a spans λ_a and F_b spans λ_b. Each surface is expressed as a union of triangles so that

$$F_a = \bigcup_{i=1}^{N_a} \Delta_{a,i} \qquad \text{and} \qquad F_b = \bigcup_{j=1}^{N_b} \Delta_{b,j}.$$

We assume that everything is in general position so that, for example, the edges of F_a meet F_b only in the interiors of the triangles $\Delta_{b,j}$.

A little geometry shows that the statement of the proposition is true for individual triangles: $|\Delta_{a,i} \cap \partial\Delta_{b,j}|^* = |\partial\Delta_{a,i} \cap \Delta_{b,j}|^*$ — see Figure 5.20. We want to break down the intersection number of the whole surface and to sum up the contributions made by all the triangles in turn.

When a surface is orientable, all its triangles can be oriented coherently so that whenever two triangles have an edge in common, the edge is oriented in opposite directions in the two triangles. This means that the edge is included twice in the sum of intersection numbers, once with each sign, making a zero net contribution. Hence, the only non-zero contributions come from edges in the boundary of the surface. Therefore,

$$
\begin{aligned}
|F_a \cap \lambda_b|^* &= \sum_{i=1}^{N_a} |\Delta_{a,i} \cap \lambda_b|^* \\
&= \sum_{i=1}^{N_a} \sum_{j=1}^{N_b} |\Delta_{a,i} \cap \partial\Delta_{b,j}|^* \\
&= \sum_{i=1}^{N_a} \sum_{j=1}^{N_b} |\partial\Delta_{a,i} \cap \Delta_{b,j}|^*
\end{aligned}
$$

$$
\begin{aligned}
&= \sum_{j=1}^{N_b} |\lambda_a \cap \Delta_{b,j}|^* \\
&= |\lambda_a \cap F_b|^*.
\end{aligned}
$$

$\square$

Theorem 5.7.3. *The sum of the intersection numbers of a simple loop λ with an orientable surface F is the linking number:* $|F \cap \lambda|^* = \mathrm{lk}(\partial F, \lambda)$.

PROOF. By Lemma 5.7.1, it is sufficient to verify the statement for a projection surface. This is left as an exercise (Exercise 5.9.10). $\square$

The following corollary will be useful later.

Corollary 5.7.4. *If λ is a simple loop that is disjoint from an oriented surface F then* $\mathrm{lk}(\lambda, \partial F) = 0$.

5.8 Genus of satellites

The behaviour under factorisation is always one of the first questions to be studied when a new link invariant is discovered. We have seen that link factorisation often produces simple relationships between the invariants of a product and its factors. The invariants discussed so far give examples of both explicit formulae (genus, bridge index) and inequalities (polygon index). Even in the cases where no relationship is proved, there is usually a conjecture about expected behaviour, sometimes backed up by partial results.

In contrast, the effect of the satellite construction on link invariants is, in general, poorly understood. A construction may provide an upper bound but lower bounds are hard to come by. Genus is unusual in this respect.

Theorem 5.8.1. *The genus of a satellite knot S constructed from pattern P with companion C, framing zero, and winding number n is bounded below:*

$$
g(S) \geqslant g(P) + n \, g(C).
$$

PROOF. If C is the trivial knot then $S = P$ and there is nothing to prove. So we assume that C is non-trivial.

Let F be a minimal-genus surface spanning S, and let V be the companion solid torus associated with C. We shall manipulate F step by step until it is easy to see how it can be related to surfaces spanning C and P.

Assume that F and ∂V are in general position. Since $|\partial F \cap \partial V| = \varnothing$, the two surfaces intersect in a set of loops.

STEP 1. The loops do not bound discs in ∂V.

Suppose $\lambda \in F \cap \partial V$ is a loop which is innermost on ∂V, and bounds a disc $\Delta \subset \partial V$ such that $\Delta \cap F = \partial \Delta = \lambda$. Then Δ is a compressing disc for F and the standard argument shows that $|F \cap \partial V|$ can be reduced.

Hence, we can assume that all loops of $F \cap \partial V$ are parallel on ∂V and cut it into a set of annuli. Number the loops $\lambda_1, \ldots, \lambda_r$ in their natural sequence around ∂V, and give them the orientations induced from the orientation of F.

STEP 2. All loops have the same orientation on ∂V.

Suppose that two adjacent loops, λ_i and λ_{i+1}, have opposite orientations. Cut F along the two loops, and glue in two copies of the annulus in ∂V bounded by λ_i and λ_{i+1}. The new surface F' is still orientable, has the same Euler characteristic as F, and has fewer intersections with ∂V. There are now two cases.

1. If F' is connected then it has the same genus as F and is therefore a minimal-genus spanning surface for S. We continue the process with F', renaming it F.

2. If F' is disconnected then one component, F_b', has boundary S and the other component, F_c', is closed. Now $g(F_b') + g(F_c') = g(F) + 1$. If $g(F_b') > g(F)$ then $g(F_c') = 0$, in which case F_c' is a sphere, λ_i spans a disc in $S^3 - V$, and the companion C is the trivial knot — contrary to assumption. Therefore F_b' is a minimal-genus spanning surface for S. We discard F_c' and continue with F_b', again renaming it F.

By repeating this process if necessary, we can assume that all the loops have the same orientation.

STEP 3. There are n loops.

Consider the surface $F_V = F \cap V$. It has $r + 1$ boundary components: $\lambda_1, \ldots, \lambda_r, \partial F$. Let m be a loop that runs parallel to a meridian of ∂V but which lies just outside V. Then m is a loop that does not meet F_V, and hence, by Corollary 5.7.4, $\mathrm{lk}(m, \partial F_V) = 0$. This implies

$$\pm r = \mathrm{lk}(m, \lambda_1 \cup \cdots \cup \lambda_r) = -\mathrm{lk}(m, \partial F) = -\mathrm{lk}(m, S) = \pm n.$$

STEP 4. $F \cap (S^3 - V)$ has n components.

Suppose that a component F^* of $F \cap (S^3 - V)$ has more than one boundary component. That is, suppose that there are two loops λ_i and λ_j that are both in ∂F^* and are such that $\lambda_k \cap F^* = \varnothing$ for all $i < k < j$.

Choose a path $\alpha_V \subset \partial V$ that connects λ_i to λ_j. Choose a path $\alpha_F \subset F^*$ that has the same endpoints as α_V. These two arcs together form a loop, $A = \alpha_F \cup \alpha_V$. We now lift this loop. Take α_V^+ to be α_V displaced so that it runs just outside V, and take α_F^+ to be α_F lifted to lie just above F^*. Then $A^+ = \alpha_F^+ \cup \alpha_V^+$ is a loop that punctures F^* once. Hence, by Theorem 5.7.3, the loop A^+ has linking number ± 1 with the boundary of F^*.

Now, A^+ lies outside V and hence it will wind around V. Suppose that $\mathrm{lk}(A^+, \lambda_i) = k$ for some (possibly zero) $k \in \mathbb{Z}$. Since $\partial F^* \subset \partial V$, the loop A^+ has linking number k with each boundary component of F^*. Therefore, $\mathrm{lk}(A^+, \partial F^*) = k \, |\partial F^*|$. But this can never be ± 1 since $|\partial F^*| \geqslant 2$ and $k \in \mathbb{Z}$, so we obtain a contradiction.

STEP 5. The loops are preferred longitudes of ∂V.

Consider the loop λ_i and the surface F_i in $S^3 - V$ which it bounds. Let C be the companion knot sitting inside V. Since F_i and C lie on opposite sides of ∂V, they do not intersect and (Corollary 5.7.4 again) $\mathrm{lk}(\lambda_i, C) = 0$. Hence λ_i is a preferred longitude of ∂V.

STEP 6. Conclusion.

Let h be the faithful homeomorphism that maps the pattern $P \subset W$ to $S \subset V$. Recall that $F_V = F \cap V$. The preimage $h^{-1}(F_V)$ is a surface in the unknotted solid torus W. The preimages $h^{-1}(\lambda_i)$ of the loops are preferred longitudes on ∂W. These can be attached to n discs to form a surface, F_P, that spans P.

Let F_i be the component of $F \cap (S^3 - V)$ that is bounded by λ_i. Each F_i is a surface spanning the companion C. So

$$g(S) = g(F) = g(F_P) + \sum_{i=1}^{n} g(F_i) \geqslant g(P) + n \, g(C).$$

$\square$

5.9 Exercises

1. Embeddings of surfaces that look different at first sight may be the same. Show that the surface shown here is isotopic to the one in Figure 5.5.

2. Show that double knots have genus 1.

3. Draw projection surfaces for pretzel knots. Show that the surface has genus 1 when all of p, q and r are odd. What happens when one of the coefficients is even?

4. Draw projection surfaces for the twist knots $C(-n, -2)$. Consider the cases when n is odd and n is even, and show that both give rise to genus-1 projection surfaces.

5. Show that, for each n, there is a pretzel presentation of $C(-n, -2)$ which gives a genus-1 projection surface. (Hint: use Exercise 3.10.12.)

6. The genus-1 knots with at most nine crossings are 3_1, 4_1, 5_2, 6_1, 7_2, 7_4, 8_1, 8_3, 9_2, 9_5, 9_{35}, 9_{46} (diagrams in Appendix A). Show that each of them is a pretzel knot, all of whose coefficients are odd.

7. Show that various choices of sign in $P(\pm 3, \pm 3, \pm 2)$ give the knots 8_5, 8_{19}, 8_{20}, and their mirror images.

8. Show that the Kinoshita–Terasaka knot (Figure 4.12) is a ribbon knot.

9. Show that $K \# K^*$ is a ribbon knot for any knot K.

10. Complete the proof of Theorem 5.7.3.

11. Show that the (p, q) torus knot is spanned by a projection surface of genus $\frac{1}{2}(p-1)(q-1)$ when $p > q \geqslant 2$.

12. Show that changing the orientation of some components in a link diagram can affect the genus of the resulting projection surface.

13. Using the diagram below, show that the double of the trefoil has a projection surface of genus 3, whatever the framing [118].

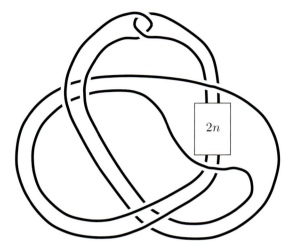

14. Show that a diagram whose Seifert graph is a tree must be a diagram of the trivial knot. Find an example to show that the converse is false.

15. Let D be the diagram of a Lissajous knot with coprime frequencies $B_y > B_x > 1$, viewed along the z-axis. Show that the Seifert graph constructed from D is a rectangular grid with vertices arranged in B_x rows and B_y columns. Hence D has $2B_x B_y - B_x - B_y$ crossings, and $B_x B_y$ type I Seifert circles.

16. Suppose that a non-split link $L = K_1 \cup \cdots \cup K_n$ has a disconnected spanning surface $F = F_1 \cup \cdots \cup F_n$ with n components such that $\partial F_i = K_i$: the components of L bound disjoint surfaces. Such a link is called a *boundary link*.

(a) Show that $\mathrm{lk}(K_i, K_j) = 0$ when $i \neq j$.

(b) Find examples of boundary links.

6 Matrix Invariants

In this chapter we introduce some invariants that are more sophisticated. In comparison with the 'minimise a geometrical property' type of invariant, they are more difficult to define. However, they have the great advantage that they can be directly calculated. They are powerful enough to distinguish many of the knots with at most nine crossings: we can finally prove that non-trivial knots do exist, and also that some knots are cheiral. We can even find a computable lower bound on the unknotting number, which allows the unknotting numbers of some knots to be established.

We start by analysing the set of loops on a surface. With an appropriate rule of composition, these form a group called the first homology group of the surface. By studying the ways these loops are arranged in space, we obtain a matrix which is dependent on the embedding of the surface. This then leads to numerical invariants of the surface boundary. (The chapter title is something of a misnomer since the matrix itself is not a link invariant, only the numbers derived from it.) Readers familiar with homology theory can skip to §6.5.

6.1 Loops in graphs

Let G be a connected oriented graph with vertex set V and edge set E. Each edge $e \in E$ is an ordered pair $[v_i, v_j]$: the edge is oriented from v_i to v_j.

We shall consider the set of all formal linear combinations of edges with coefficients taken from the abelian group $(\mathbb{Z}, +)$. Thus a typical element can be written

$$c = \sum_{i=1}^{n} \lambda_i \, e_i$$

where n is the number of edges of G, and $\lambda_i \in \mathbb{Z}$. An expression of this form is called a *1-chain*.

Any path or circuit in G has a corresponding 1-chain: the coefficient λ_i is the number of times the path follows edge e_i, counted with sign: we use the convention that $-e$ denotes the edge e with the reverse orientation. If the path is simple, all the λ_i will be 0 or ± 1. In general, a chain need not correspond to a path or circuit in a natural way.

Two chains can be combined according to the following rule:

$$\sum_{i=1}^{n} \lambda_i\, e_i \;+\; \sum_{i=1}^{n} \mu_i\, e_i \;=\; \sum_{i=1}^{n} (\lambda_i + \mu_i)\, e_i.$$

With this operation, the set of 1-chains becomes an abelian group, denoted by $C_1(G)$, isomorphic to $\mathbb{Z} \oplus \cdots \oplus \mathbb{Z} = \mathbb{Z}^n$.

We can follow a similar procedure and define the group of 0-chains, $C_0(G)$, based on formal linear combinations of the vertices of G. Vertices cannot be oriented so there is no convention about $-v$.

There is a group homomorphism $\partial : C_1(G) \longrightarrow C_0(G)$ known as the boundary operator: $\partial([v_i, v_j]) = v_j - v_i$. The map is linear so we get

$$\partial\left(\sum_{i=1}^{n} \lambda_i\, e_i \right) \;=\; \sum_{i=1}^{n} \lambda_i\, \partial(e_i).$$

The kernel of this map is a special subgroup of $C_1(G)$: the 1-chains that map to zero have no boundary and are called 1-*cycles*. Check that the 1-chain corresponding to a circuit in G is a 1-cycle. The subgroup of 1-cycles is denoted by $Z_1(G)$ — from the German word Zyklus. The element $0 \in Z_1(G)$ is called the *trivial* cycle.

Theorem 6.1.1.   *A tree has no non-trivial 1-cycles.*

PROOF. Let z be a 1-cycle in a tree G. The proof proceeds by induction on n, the number of edges in G.

If $n = 1$ then

$$0 \;=\; \partial(z) \;=\; \lambda_1\, \partial(e_1) \;=\; \lambda_1\, v_2 - \lambda_1\, v_1.$$

There are no other edges to contribute extra terms, hence $\lambda_1 = 0$.

Suppose now that $n > 1$ and that edge e_n is a leaf connecting vertices v_1 and v_2 where v_1 is the terminal node. Then

$$0 \;=\; \partial(z) \;=\; \partial\left(\sum_{i=1}^{n} \lambda_i\, e_i \right) \;=\; \partial\left(\sum_{i=1}^{n-1} \lambda_i\, e_i \right) + \lambda_n\, \partial(e_n).$$

We also have $\pm\partial(e_n) = v_2 - v_1$. Since v_1 is a terminal vertex, there are no other terms involving it and we deduce that $\lambda_n = 0$. Hence z is a cycle on the tree $G - e_n$, which has $n - 1$ edges. The result follows by induction. $\qquad\square$

A *basis* for $Z_1(G)$ is a set of cycles such that each 1-cycle in $Z_1(G)$ can be expressed uniquely as a linear combination of the basis elements. It is in this sense that a graph has a maximum number of independent circuits.

We can construct a basis for $Z_1(G)$ as follows.

1. Let T be a spanning tree for G.

2. Label the edges of G so that the first r edges are not in the tree: $e_1, \ldots, e_r \notin T$, and $e_{r+1}, \ldots, e_n \in T$.

3. For each edge e_i not in T, the graph $T \cup e_i$ contains a unique circuit: let $z_i \in Z_1(G)$ be the 1-cycle corresponding to this circuit with the orientation chosen so that the coefficient of e_i is $+1$. This gives a set of r 1-cycles.

Note that r is the rank of the graph and is independent of the tree chosen.

Theorem 6.1.2. *The set of 1-cycles just constructed forms a basis for* $Z_1(G)$.

PROOF. If G is a tree, $r = 0$ and the basis is empty. By the previous theorem, there are no non-trivial 1-cycles, so the theorem is true in this case. Now assume that G is not a tree and $r > 0$.

Suppose that $z \in Z_1(G)$ is a 1-cycle. We need to find coefficients $\nu_i \in \mathbb{Z}$ such that

$$z = \sum_{i=1}^{r} \nu_i \, z_i.$$

Since z is a 1-cycle, it is also a 1-chain and we can write it as

$$z = \sum_{i=1}^{n} \lambda_i \, e_i.$$

The first r of these λ_i are the required coefficients. To see this we show that the difference is zero. Consider

$$z - \sum_{i=1}^{r} \lambda_i \, z_i.$$

This expression is a cycle since it is a sum of cycles. As a 1-chain, it has coefficient zero on edges $e_1, \ldots, e_r$ by construction, so it is a 1-cycle in the tree T. A tree has no non-trivial 1-cycles, hence the coefficients on edges $e_{r+1}, \ldots, e_n$ are also zero.

It remains to show that the coefficients are unique (Exercise 6.9.2).   □

6.2 Simplicial complexes

We are now going to generalise the discussion in the previous section to two dimensions. This will enable us to create an algebraic structure for a set of loops in a surface.

Graphs are essentially 1-dimensional objects, being composed of lines and points. We know that all surfaces can be built up from triangles, so we add these to our collection of building blocks. These elements — point (vertex), line (edge), triangle — form the start of the simplex family. There is one simplex in each dimension; the next in the sequence is the solid tetrahedron.

The standard n-dimensional *simplex* is the subset of $\mathbb{R}^{n+1}$ given by

$$\left\{ (x_1, \ldots, x_{n+1}) \in \mathbb{R}^{n+1} : \sum_{i=1}^{n+1} x_i = 1 \right\}.$$

Thus the standard 0-simplex is the point $1 \in \mathbb{R}$, the standard 1-simplex is the line segment in $\mathbb{R}^2$ that connects the points $(1,0)$ and $(0,1)$, and the standard 2-simplex is the equilateral triangle in $\mathbb{R}^3$ with vertices at $(1,0,0)$, $(0,1,0)$, and $(0,0,1)$. When simplices are used as building blocks, we allow these standard forms to be transformed by piecewise linear homeomorphisms: to assemble a simplicial structure we can use any straight line segment, and any triangle.

A simplex is bounded by simplices of lower dimension and these are called its *faces*. Thus an edge (1-simplex) has two vertices (0-simplices) as faces. A triangle (2-simplex) has six faces: three edges and three vertices.

A *simplicial complex* X is a finite collection of (homeomorphic copies of) simplices glued together so that the following conditions are satisfied:

- if $\sigma \in X$ then any face of σ is also in X,
- if σ and σ' are in X then $\sigma \cap \sigma'$ is either empty or a face of both σ and σ'.

Examples include graphs, triangulated surfaces, and the hybrid object shown in Figure 6.1.

6.3 Loops in simplicial complexes

Given a simplicial complex, we can consider the set of all formal linear combinations of oriented simplices with coefficients taken from an abelian group G. We usually take G to be the set of integers under addition $(\mathbb{Z}, +)$. An n-dimensional *chain* in a simplicial complex X is a linear combination of oriented n-simplices. There is only one rewriting rule: if σ and σ' denote the same simplex with opposite orientations then $\sigma + \sigma' = 0$.

The set of n-chains forms a group $C_n(X)$: composition of chains is defined by componentwise addition of the values on each simplex.

A chain is *elementary* if its value is 1 on exactly one simplex, and zero elsewhere.

Example 6.3.1. We shall work with the simplicial complex X shown in Figure 6.1. It is composed of six vertices, seven edges, and one triangle. The edges and triangle are oriented as shown. The edge or vertex labelled i will be denoted by e_i or v_i, respectively. A general 1-chain can be written as a sum of elementary 1-chains. By abusing notation we can write e_i for both the oriented edge (1-simplex) and the elementary 1-chain derived from it. Then a general 1-chain c can be written

$$c = \sum_{i=1}^{7} \lambda_i e_i \quad \text{where } \lambda_i \in G.$$

Thus $C_1(X) = G \oplus \cdots \oplus G = G^7$. Similarly, $C_0(X) = G^6$ and $C_2(X) = G$. □

By generalising this example we see that:

Theorem 6.3.2. *For $G = (\mathbb{Z}, +)$ the group $C_n(X)$ is free abelian and is generated by the elementary chains obtained by giving each n-simplex an orientation.*

Note that there is no natural basis for $C_n(X)$: when $n > 0$ the choice of orientation of each n-simplex is arbitrary; vertices (the case $n = 0$) cannot be oriented.

We now define two boundary operators. We write $[v_i, v_j]$ for the edge connecting vertex v_i to v_j oriented from v_i to v_j, and $[v_i, v_j, v_k]$ for the 2-simplex

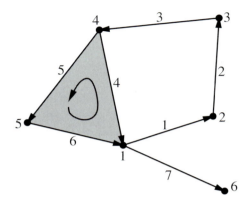

Figure 6.1. A simplicial complex.

spanned by the given vertices and oriented by traversing the boundary from v_i to v_j to v_k. The operators ∂_1 and ∂_2 are defined below on individual simplices and extended to chains by linearity:

$$\partial_1 : C_1(X) \longrightarrow C_0(X)$$
$$[v_1, v_2] \longmapsto v_2 - v_1;$$

$$\partial_2 : C_2(X) \longrightarrow C_1(X)$$
$$[v_1, v_2, v_3] \longmapsto [v_1, v_2] + [v_2, v_3] + [v_3, v_1].$$

The kernel of ∂_1 consists of those 1-chains which have zero boundary. These elements of $C_1(X)$ are called 1-cycles, and the set of them is denoted by $Z_1(X)$.

$$Z_1(X) = \ker(\partial_1) = \{c \in C_1(X) : \partial_1(c) = 0\} \subset C_1(X).$$

Some 1-chains are boundaries of 2-chains. Denote the set of these boundaries by $B_1(X)$. Then

$$B_1(X) = \text{Im}(\partial_2) = \{c \in C_1(X) : \exists\, t \in C_2(X) \text{ with } \partial_2(t) = c\} \subset C_1(X).$$

In fact, such boundaries are always cycles so $B_1(X) \subset Z_1(X)$. Verify this by showing that $\partial_1 \circ \partial_2 = 0$ for all 2-chains (Exercise 6.9.3).

Example 6.3.3. We continue analysing the simplicial complex in Figure 6.1. The boundary of a general 1-chain is

$$\partial_1(c) = \partial_1 \left(\sum_{i=1}^{7} \lambda_i\, e_i \right)$$

$$= \sum_{i=1}^{7} \lambda_i \, \partial_1(e_i)$$

$$= (\lambda_4 + \lambda_6 - \lambda_1 - \lambda_7)\, v_1 \; + \; (\lambda_1 - \lambda_2)\, v_2 \; + \; (\lambda_2 - \lambda_3)\, v_3$$
$$+ \; (\lambda_3 - \lambda_4 - \lambda_5)\, v_4 \; + \; (\lambda_5 - \lambda_6)\, v_5 \; + \; \lambda_7\, v_6.$$

For this to be a cycle, the value on each elementary 0-chain (vertex) must be zero. This immediately gives us that $\lambda_7 = 0$, and other simple relations between the coefficients: $\lambda_1 = \lambda_2 = \lambda_3$, and $\lambda_5 = \lambda_6$. We can further deduce $\lambda_4 = \lambda_3 - \lambda_5$. Thus once λ_3 and λ_5 are known, all the other coefficients can be determined. Putting (λ_3, λ_5) as $(1,0)$ and $(0,1)$ shows that $Z_1(X)$ is the free abelian group $G \oplus G$ generated by $e_1 + e_2 + e_3 + e_4$ and $e_5 + e_6 - e_4$.

There is only one 2-simplex and its boundary is $e_5 + e_6 - e_4$. Hence $B_1(X)$ is just G generated by this 1-cycle.   □

The sets $B_1(X)$ and $Z_1(X)$ are both closed under composition so they are both subgroups of $C_1(X)$. Consider the sequence of homomorphisms

$$C_2(X) \xrightarrow{\partial_2} C_1(X) \xrightarrow{\partial_1} C_0(X)$$

and note the fact that $C_1(X)$ is abelian implies that $B_1(X)$ is a normal subgroup of $Z_1(X)$. Hence we can form the quotient group:

$$\ker(\partial_1)/\mathrm{Im}(\partial_2) \; = \; Z_1(X)/B_1(X) \; = \; H_1(X; G).$$

This group is called the first homology group of X. Often, the coefficient group is omitted and we just write $H_1(X)$.

The output of the quotient process can also be obtained by defining an equivalence relation, $\sim$, on the set of 1-cycles. Two 1-cycles are equivalent, or *homologous*, if they bound a 2-chain: that is $z_1 \sim z_2$ if and only if there is some 2-chain t such that $z_1 - z_2 = \partial_2(t)$. A 1-cycle that is a 1-boundary is said to be homologous to zero: $z \sim 0$ for all $z \in B_1(X)$.

Example 6.3.4. The homology group of the space X shown in Figure 6.1 is

$$H_1(X) \; = \; Z_1(X)/B_1(X) \; = \; (G \oplus G)/G \; = \; G.$$

The boundary cycle $e_5 + e_6 - e_4$ is quotiented out and $H_1(X)$ is generated by the cycle $e_1 + e_2 + e_3 + e_4$. Geometrically, we can think of the homology as having detected the 'hole' in the complex.   □

So the cycle $e_1 + e_2 + e_3 + e_4$ contains all the essential information about the loops in the complex X. From the homological viewpoint, the rest of the complex is redundant. In fact, there is a method for deleting simplices from a simplicial complex without changing its homology. We shall look at two examples.

Consider first the edge e_7 and the vertex v_6. This vertex is a 'terminal' vertex: it is not the endpoint of any edge other than e_7. Let z be a 1-cycle in X so that

$$0 = \partial(z) = \sum_{i=1}^{n} \lambda_i \, \partial(e_i).$$

Now $\partial(e_7) = v_6 - v_1$. Since v_6 is not a face of any other edge, we must have that $\lambda_7 = 0$. Hence e_7 always has coefficient zero on any cycle, and can be deleted from X without affecting the homology. To delete e_7 we must also delete v_6.

Now we consider the 2-simplex t and the edge e_6. Suppose we have a 1-cycle $z \in X$ which has value λ_6 on e_6. Form a new 1-cycle $z' = z - \lambda_6 \, \partial t$. Adding 1-cycles produces another 1-cycle. Adding a boundary to a 1-cycle does not affect its homology class. Therefore z' and z represent the same element in $H_1(X)$. Now the coefficient of e_6 in z' is zero, so we can delete e_6 without affecting the homology. To delete e_6 we must also delete t.

This discussion can be formalised as follows.

A simplex is *principal* if it is not a face of any simplex. A simplex is *free* if it is a face of exactly one simplex. Thus a free face must be the face of a principal simplex. In the example we have been studying, the principal faces are e_1, e_2, e_3, e_7 and t; the free faces are v_6 of e_7, and e_4, e_5 and e_6 of t.

If τ is a free face of a principal face σ in a simplicial complex X then we say that the simplicial complex $X - (\sigma \cup \tau)$ is obtained by *collapsing* τ across σ. The process is called an *elementary collapse*. If X_1 can be transformed into X_2 by a finite sequence of elementary collapses, we say that X_1 collapses to X_2.

The importance of this idea comes from the following theorem.

Theorem 6.3.5.   *If a simplicial complex X_1 collapses to a simplicial complex X_2 then their homology groups are isomorphic: $H_1(X_1) \cong H_1(X_2)$. Furthermore, 1-cycles in X_2 whose homology classes generate $H_1(X_2)$ have homology classes in X_1 which generate $H_1(X_1)$.*

This is a very brief introduction to part of the powerful machinery of algebraic topology. It covers only the aspects of homology theory we need. All of the discussion generalises to give H_n in the obvious way — the group $H_n(X)$ counts the number of independent n-dimensional 'holes' in X. Homology groups can be used to distinguish different topological objects but we shall not pursue this. We are interested only in the sets of generators of H_1 for surfaces. It is these that we examine next.

6.4 Homology of surfaces

We have observed before (Fact 2.7.1) that a surface can be triangulated. This means that we can find a simplicial complex that is homeomorphic to the given surface.

We regard the following theorem in the same way as the facts summarised in Chapter 2.

Fact 6.4.1. *For a surface F, the homology group $H_1(F)$ is independent of the triangulation chosen.*

A loop in a triangulated surface F is isotopic (hence homologous) to a 1-chain: it can be deformed to lie in the edge network. We can now identify characteristics of the surface that are related to its homology.

Any loop that bounds a disc in the surface is homologically trivial (because it is a boundary). In fact, any separating loop in a *closed* surface is a boundary (it bounds each of the 2-chains produced by the separation) so it is homologous to zero. The loop around the 'waist' of a pretzel is an example of a trivial 1-cycle that does not bound a disc.

Non-separating curves in F do not bound and must be non-trivial elements of $H_1(F)$. Recall that we can think of a connected surface of genus g as a sphere with g handles attached and possibly some discs removed. Each handle contributes a pair of non-separating loops (see Figure 6.2). For a closed surface, these loops are the only non-trivial 1-cycles. For surfaces with boundary, there can be separating curves which do not bound (the core of an annulus for example): these are also non-trivial homology elements.

A triangulation of a compact surface F has a finite number of vertices, edges and triangles. This implies that $H_1(F)$ is finitely generated. Figure 6.3 shows a set of loops forming a basis for H_1 of the generic disc–band surface of Figure 5.14. Each handle in the surface contributes a pair of bands at the top of the disc, which contain the two loops shown in Figure 6.2.

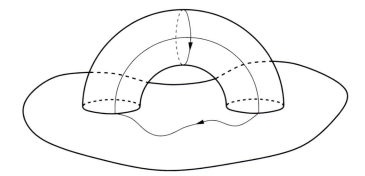

Figure 6.2. Each handle on a surface contributes two non-separating loops.

There is also a loop for each boundary component after the first one. Thus there are $2g + \mu - 1$ loops in total: notice that this is $1 - \chi(F)$.

A surface spanning a link is a surface with a boundary. The edges along the boundary are free edges and can be collapsed, so removing triangles from the rim of the surface. This process can be continued until all the triangles have been deleted leaving just a graph.

Theorem 6.4.2. *Let F be a projection surface constructed from a link diagram, and let G be its Seifert graph. Then F collapses to G.*

Corollary 6.4.3. *The homology groups of a projection surface and its Seifert graph are isomorphic: $H_1(F) \cong H_1(G)$.*

This means that we can apply the method of Theorem 6.1.2 to find a homology basis for a projection surface F:

1. Construct the Seifert graph G for the projection surface F by placing a vertex in each disc and an edge in each band.

2. Choose a spanning tree T for G.

3. For each edge $e_i \in (G - T)$, the graph $T \cup e_i$ contains a unique circuit: let z_i be the 1-cycle corresponding to this circuit. The set of these 1-cycles is a homology basis for F.

This is the 'algebra of loops in a surface' that we set out to develop. We can now apply it to produce some powerful link invariants.

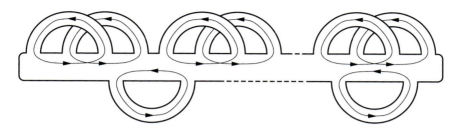

Figure 6.3. A homology basis for a generic disc–band surface.

6.5 The Seifert matrix

In this section we shall explore a technique for studying the embeddings of connected oriented surfaces in $\mathbb{R}^3$.

First we thicken the surface as follows. Recall that an oriented surface has positive and negative sides. Let $b : F \times [-1, 1] \longrightarrow \mathbb{R}^3$ be a homeomorphism such that $b(F \times \{0\}) = F$ and $b(F \times \{1\})$ lies on the positive side of F. Any subset $X \subset F$ can be lifted out of the surface on either side: we let $X^+ = b(X \times \{1\})$ and $X^- = b(X \times \{-1\})$.

We are interested in the case that X is a loop in the surface. Two loops in the surface may intersect, but if we lift one of them out of the surface this is no longer possible; furthermore the two loops then have a linking number. We can use this to define the following mapping:

$$\Theta : H_1(F) \times H_1(F) \longrightarrow \mathbb{Z}$$
$$(a, b) \longmapsto \text{lk}(a, b^+).$$

This is called the *Seifert pairing* or *linking form* of the embedded surface.

Example 6.5.1. We shall compute the linking form for the projection surface F constructed from the diagram of 10_{165} shown in Figure 5.3. The surface has been redrawn in Figure 6.4 and six loops lying in the surface have been added. These loops are the boundaries of the faces of the Seifert graph and they form a basis for $H_1(F)$.

The cut vertex in the Seifert graph corresponds to the inner of the two octagonal Seifert circles. This type II Seifert circle divides the surface naturally into upper and lower parts. The loops a, b and c lie in the lower part, and the loops d, e and f lie in the upper part. Assume that F is oriented in such a way that the upper surface of the disc spanning the outermost

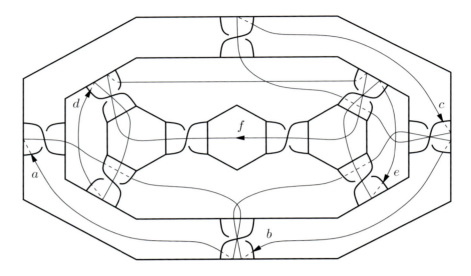

Figure 6.4. A projection surface for 10_{165} — the set of loops forms a basis for $H_1(F)$.

Seifert circle is positive. Then the linking form is summarised in the following table, called a *Seifert matrix*:

	a^+	b^+	c^+	d^+	e^+	f^+
a	1	-1	0	0	0	0
b	0	0	0	0	0	0
c	0	1	0	0	0	0
d	1	0	0	-1	0	1
e	0	-1	1	0	1	-1
f	0	0	-1	0	0	-1

Let us now examine how the entries in the table are calculated. Since the projections of loops d and e are disjoint, $\mathrm{lk}(d, e+) = \mathrm{lk}(e, d+) = 0$. This also applies to pairs (a, c), (a, e), (a, f), (b, d), (b, f) and (c, d).

The block of zeroes in the top-right is a consequence of the division of the surface into upper and lower parts: d^+, e^+ and f^+ come from loops in the upper part and they are lifted away from the loops a, b and c in the lower part, hence the two sets do not link.

Now let us turn to the more interesting cases. Most of them are shown in Figure 6.5. Consider first the pair a and d shown in the top-left. There is a band that loop a passes underneath and loop d passes through. This gives rise to the crossing in the figure. The two loops also have a point

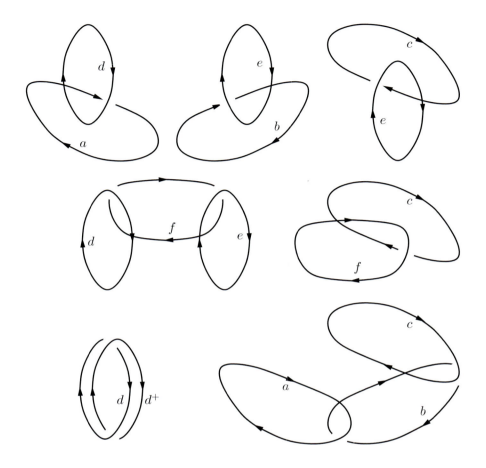

Figure 6.5.

of intersection. Lifting loop d to form d^+ converts this intersection into a crossing of d^+ over a and we see that $\mathrm{lk}(a, d+) = 0$. Lifting loop a converts the intersection into a crossing of a^+ over d. These two loops have non-zero linking number: $\mathrm{lk}(d, a+) = 1$. All the other pairs of distinct loops are shown in Figure 6.5 and similar analyses yield all the off-diagonal entries in the table.

The remaining cases are self-linking forms. The case $\mathrm{lk}(d, d+)$ is shown on the bottom-left of Figure 6.5. The value is half the sum of the signs of all the crossings that d passes through. $\qquad\square$

This example has produced a Seifert matrix M that depends on the surface F, its embedding in $\mathbb{R}^3$, and the choice of basis for $H_1(F)$. A change of basis is reflected in the Seifert matrix by a congruence transformation of the form $M \longrightarrow P^{\mathrm{T}} M P$ where $P \in \mathrm{GL}(6, \mathbb{Z})$ is an invertible matrix with integer entries.

Although we used a projection surface as an example, the construction works for orientable surfaces in general.

6.6 Link invariants from the Seifert matrix

The Seifert matrix itself is not a link invariant. A link can be spanned by surfaces with different genera, and even sometimes by non-isotopic embeddings with the same genus. However, we know (Theorem 5.5.1) that any two surfaces spanning a given link are related by a sequence of piping and compressing operations. If we derive a feature of the matrix which is invariant under these operations then we have found an invariant of the surface boundary.

Lemma 6.6.1. *Let M be a Seifert matrix of a connected surface F, and let $\widehat{F}$ be a surface derived from F by adding a tube. Then there is a basis for $H_1(\widehat{F})$ which contains the basis elements of $H_1(F)$ as a subset, and such that the Seifert matrix of $\widehat{F}$ has the form*

$$
\begin{pmatrix}
 & & & * & 0 \\
 & M & & \vdots & \vdots \\
 & & & * & 0 \\
0 & \cdots & 0 & 0 & 1 \\
0 & \cdots & 0 & 0 & 0
\end{pmatrix}
\quad or \quad
\begin{pmatrix}
 & & & 0 & 0 \\
 & M & & \vdots & \vdots \\
 & & & 0 & 0 \\
* & \cdots & * & 0 & 0 \\
0 & \cdots & 0 & 1 & 0
\end{pmatrix}
$$

where the asterisks indicate unknown integers.

PROOF. Suppose we have added a tube to the surface as shown in Figure 6.2. Let m be the meridian of the tube and let l be another curve that runs along the tube and back through F. Assume the outside of the tube is the positive side of the surface.

We can see that $\mathrm{lk}(m, l^+) = 0 = \mathrm{lk}(m, m^+)$ and $\mathrm{lk}(l, m^+) = 1$. We can also choose l so that $\mathrm{lk}(l, l^+) = 0$ for if $\mathrm{lk}(l, l^+) = \lambda \neq 0$ we can replace l with $l - \lambda m$:

$$
\begin{aligned}
& \mathrm{lk}((l - \lambda m), (l - \lambda m)^+) \\
={} & \mathrm{lk}(l, l^+) - \lambda \, \mathrm{lk}(m, l^+) - \lambda \, \mathrm{lk}(l, m^+) + \lambda^2 \, \mathrm{lk}(m, m^+) \\
={} & \mathrm{lk}(l, l^+) - \lambda \\
={} & 0.
\end{aligned}
$$

Let $(a_1, \ldots, a_n)$ be a basis for $H_1(F)$. Clearly, $\mathrm{lk}(a_i, m^+) = \mathrm{lk}(m, a_i^+) = 0$ for all i. The added tube may be knotted or linked around other parts of the surface so we have no control over the ways that a_i^+ links with l or a_i links with l^+. Let $\lambda_i = \mathrm{lk}(l, a_i^+)$. Then adding a tube to F corresponds to an enlargement of the matrix M of the form

$$
\begin{array}{c|ccccc}
 & a_1^+ & \cdots & a_n^+ & l^+ & m^+ \\
\hline
a_1 & & & & * & 0 \\
\vdots & & M & & \vdots & \vdots \\
a_n & & & & * & 0 \\
l & \lambda_1 & \cdots & \lambda_n & 0 & 1 \\
m & 0 & \cdots & 0 & 0 & 0
\end{array}
$$

Using the same trick as above, we shall change the basis of $H_1(\widehat{F})$ so that the λ_i become zero. Replace each generator a_i with $b_i = a_i - \lambda_i m$. Then

$$
\mathrm{lk}(l, b_i^+) = \mathrm{lk}(l, (a_i - \lambda_i m)^+) = \mathrm{lk}(l, a_i^+) - \lambda_i \, \mathrm{lk}(l, m^+) = 0.
$$

As before, $\mathrm{lk}(b_i, m^+) = \mathrm{lk}(m_i, b^+) = \mathrm{lk}(b_i, l^+) = 0$. Also

$$
\begin{aligned}
\mathrm{lk}(b_i, b_j^+) &= \mathrm{lk}((a_i - \lambda_i m), (a_j - \lambda_j m)^+) \\
&= \mathrm{lk}(a_i, a_j^+) - \lambda_i \, \mathrm{lk}(m, a_j^+) - \lambda_j \, \mathrm{lk}(a_i, m^+) + \lambda_i \lambda_j \, \mathrm{lk}(m, m^+) \\
&= \mathrm{lk}(a_i, a_j^+).
\end{aligned}
$$

This last result can be seen another way. If we slide the loop m along the tube until it comes off the end then we can think of it as a loop in F. Now m bounds a disc so it is trivial in $H_1(F)$. This means that a_i and b_i are homologous in F and so the change of basis does not affect M.

The enlarged matrix now looks like the one on the left in the statement of the lemma. If the outside of the tube is the negative side of the surface, we get the other form. $\qquad\square$

Matrices which are related by a finite sequence of these enlargement and reduction operations, together with the congruence transformations $M \longrightarrow P^{\mathrm{T}} M P$ arising from a change of basis, are called *S-equivalent*. Recall that surfaces related by surgery are also said to be S-equivalent. This should not cause confusion. In any case, we have shown that:

Theorem 6.6.2. *If two surfaces are S-equivalent then their corresponding Seifert matrices are S-equivalent.*

Consequently, any property derived from a Seifert matrix that is invariant under S-equivalence is a link invariant.

Two standard numerical values associated with a symmetric matrix are its determinant and its signature.

Definition 6.6.3. The *determinant* of a link L, written $\det(L)$, is the absolute value of the determinant of $M + M^{\mathrm{T}}$ where M is any Seifert matrix for L.

Definition 6.6.4. The *signature* of a link L, written $\sigma(L)$, is the signature of $M + M^{\mathrm{T}}$ where M is any Seifert matrix for L.

It is not clear from these definitions that the determinant and signature of a link are independent of M, and hence well-defined. We shall prove this in Theorem 6.6.5. However, first we shall review some results in linear algebra.

Any symmetric matrix A with entries in $\mathbb{R}$ is congruent to a diagonal matrix: there is an invertible orthogonal matrix P with entries in $\mathbb{R}$ and determinant ± 1 such that $P^{\mathrm{T}}AP$ has all its non-zero entries on the diagonal. The signature of a diagonal matrix is the number of positive entries minus the number of negative entries. Two diagonal matrices are congruent if and only if they have the same numbers of positive entries, negative entries, and zeroes. More importantly for us, congruence preserves signature — a result known as Sylvester's Theorem.

Theorem 6.6.5. *Let M be a Seifert matrix constructed from a surface F spanning a link L. Then $|\det(M + M^{T})|$ and the signature of $M + M^{T}$ are link invariants which depend only on L.*

PROOF. We need to show that the determinant and signature are preserved by the congruence transformations and the enlargement operations. The first part is straightforward. For signature we rely on Sylvester's Theorem quoted above: signature is preserved by congruence. For the determinant, note that $\det(P) = \pm 1$ so

$$
\begin{aligned}
\det\left(P^{\mathrm{T}}MP + (P^{\mathrm{T}}MP)^{\mathrm{T}}\right) &= \det\left(P^{\mathrm{T}}(M + M^{\mathrm{T}})P\right) \\
&= \det(P^{\mathrm{T}})\det(M + M^{\mathrm{T}})\det(P) \\
&= \det(M + M^{\mathrm{T}}).
\end{aligned}
$$

Now we consider the enlargement operations described in Lemma 6.6.1.

An enlarged matrix has the following block structure:

$$
\left(
\begin{array}{ccc|cc}
 & & & * & 0 \\
 & M + M^{\mathrm{T}} & & \vdots & \vdots \\
 & & & * & 0 \\
\hline
* & \cdots & * & 0 & 1 \\
0 & \cdots & 0 & 1 & 0
\end{array}
\right).
$$

Expanding the (matrix) determinant using the 1's in the lower-right block, we get $-\det(M + M^{\mathrm{T}})$ so enlargement does not change the (link) determinant. To compute the signature, first note that the values shown as asterisks can be set to zero by performing congruence transformations. It is then easy to see that the signature of the matrix is the sum of the signatures of the two diagonal blocks. The 2×2 block introduced by the enlargement has signature zero (Exercise 6.9.11).   □

The table in Figure 6.6 shows the prime knots of up to nine crossings separated by the determinant and the absolute value of the signature. This is sufficient to distinguish all the knots of up to six crossings from each other but, moving on to seven crossings, we cannot distinguish 7_2 from 6_2. Unfortunately, many cells in the table contain more than one entry: there are *five* knots with signature 2 and determinant 27. However, we now know there are at least 48 distinct non-trivial knots.

6.7 Properties of signature and determinant

The determinant and signature are invariants we can calculate explicitly. The cost of this useful property is a more complex definition. It is not so apparent as it is with some of the 'minimise a geometric property' type of invariants how they behave under the usual link operations: reflection, reversal, factorisation, the satellite construction, and so on. These are investigated in the following theorems.

First, however, we cannot ignore the following result suggested by the examples in the table in Figure 6.6 (see also Exercise 6.9.15).

Theorem 6.7.1. *If K is a knot then $\det(K)$ is odd and $\sigma(K)$ is even.*

PROOF. Let F be an orientable surface spanning K and let M be its Seifert matrix.

Lemma 6.7.6 below shows that $\det(M - M^{\mathrm{T}}) = 1$ for a knot. Working modulo 2, we have $M + M^{\mathrm{T}} = M - M^{\mathrm{T}}$ so $\det(K) \equiv 1 \bmod 2$.

det	$\sigma = 0$	$\sigma = 2$	$\sigma = 4$	$\sigma = 6$	$\sigma = 8$
1	0_1				
3		3_1		8_{19}	
5	4_1		5_1		
7		$5_2, 9_{42}$		7_1	
9	$6_1, 8_{20}, 9_{46}$				9_1
11		$6_2, 7_2$			
13	$6_3, 8_1$		$7_3, 9_{43}$		
15		$7_4, 8_{21}, 9_2$			
17	$8_3, 9_{44}$		$7_5, 8_2$		
19		$7_6, 8_4$		9_3	
21	7_7		$8_5, 9_4$		
23		$8_6, 8_7, 9_5, 9_{45}$			
25	$8_8, 8_9$		9_{49}		
27		$8_{10}, 8_{11}, 9_{35}, 9_{47}, 9_{48}$		9_6	
29	$8_{12}, 8_{13}$		9_7		
31		$8_{14}, 9_8$		9_9	
33			$8_{15}, 9_{10}, 9_{11}$		
35		$8_{16}, 9_{12}$			
37	$8_{17}, 9_{14}$		$9_{13}, 9_{36}$		
39		$9_{15}, 9_{17}$		9_{16}	
41	9_{19}		$9_{18}, 9_{20}$		
43		$9_{21}, 9_{22}$			
45	$8_{18}, 9_{24}, 9_{37}$		9_{23}		
47		$9_{25}, 9_{26}$			
49	$9_{27}, 9_{41}$				
51		$9_{28}, 9_{29}$			
53	9_{30}				
55		$9_{31}, 9_{39}$			
57			9_{38}		
59		9_{32}			
61	9_{33}				
69	9_{34}				
75		9_{40}			

Figure 6.6. The prime knots up to nine crossings separated by signature and determinant.

Since the determinant is non-zero, there are no zeroes in the diagonalised form of $M + M^{\mathrm{T}}$. Furthermore, since K is a knot, there are $2\,g(F)$ rows and columns in M. Hence the signature is the difference of two even numbers or of two odd numbers.   □

Theorem 6.7.2.   *For any link L*

  (a) $\det(-L) = \det(L) = \det(L^*)$,

  (b) $\sigma(-L) = \sigma(L)$,

  (c) $\sigma(L^*) = -\sigma(L)$.

PROOF. Let D be a diagram of L, let F be the projection surface constructed from D, and let M be the Seifert matrix of F. If the orientation of L is reversed then the positive and negative sides of F are interchanged. The Seifert matrix for $-L$ is the transpose M^{T}. Hence $M + M^{\mathrm{T}}$ is unchanged.

To obtain L^* we switch the signs of all the crossings in D. This has the effect of changing the signs of all the linking numbers used to calculate the entries of M. Hence the Seifert matrix for L^* is $-M$. Since $\det(L)$ is defined as an absolute value, it is unchanged. The signature clearly changes sign.   □

Corollary 6.7.3.   *An amphicheiral link has signature zero.*

This shows, for example, that the left- and right-handed trefoils are distinct. Like almost all tests, this one is not infallible. Of the non-trivial knots up to eight crossings seven are amphicheiral (4_1, 6_3, 8_3, 8_9, 8_{12}, 8_{17} and 8_{18}), but the following cheiral knots also have zero signature: 6_1, 7_7, 8_1 and 8_8.

Theorem 6.7.4.   *If $L_1 \sqcup L_2$ is a split link then*

  (a) $\det(L_1 \sqcup L_2) = 0$,

  (b) $\sigma(L_1 \sqcup L_2) = \sigma(L_1) + \sigma(L_2)$.

PROOF. Let F_i be an orientable surface spanning L_i and let M_i be the Seifert matrix of F_i. A connected surface F spanning L can be formed by piping F_1 and F_2 together. To create a basis for $H_1(F)$ we take the union of the bases for F_1 and F_2 together with a meridian m of the piping tube. Now

$\operatorname{lk}(a_j, m^+) = \operatorname{lk}(m, a_j^+) = 0$ for any loop $a_j \in H_1(F_i)$. Thus a Seifert matrix for F has the form

$$\begin{pmatrix} M_1 & 0 & 0 \\ 0 & M_2 & 0 \\ 0 & 0 & 0 \end{pmatrix}.$$

$\square$

Theorem 6.7.5. *If a link can be factorised as $L_1 \# L_2$ then*

(a) $\det(L_1 \# L_2) = \det(L_1) \det(L_2)$,

(b) $\sigma(L_1 \# L_2) = \sigma(L_1) + \sigma(L_2)$.

PROOF. Let F_i be an orientable surface spanning L_i and let M_i be the Seifert matrix of F_i. A connected surface F spanning L can be formed by adding a rectangular disc R such that $R \cap F_i = \partial R \cap \partial F_i = \alpha_i$ is a single arc for each i, and α_1 and α_2 are opposite ends of the rectangle. Taking the union of the bases for F_1 and F_2 as the basis for F, we get a Seifert matrix for F of the form

$$\begin{pmatrix} M_1 & 0 \\ 0 & M_2 \end{pmatrix}.$$

$\square$

This result does not help in the detection of prime links. The fact that $\det(L)$ is a prime number does not force L to be a prime link because there are some non-trivial knots with determinant 1 (see Exercise 6.9.7).

Lemma 6.7.6. *Let M be a Seifert matrix for a surface F. Then*

$$\det(M - M^T) = \begin{cases} 1 & \text{if } |\partial F| = 1, \\ 0 & \text{if } |\partial F| > 1. \end{cases}$$

PROOF. Choose the basis for $H_1(F)$ shown in Figure 6.3. Each intersection of two oriented curves on F when viewed on the positive side of F can be given a sign, as shown in Figure 6.7. To complete the proof (Exercise 6.9.9) it is sufficient to show that $\operatorname{lk}(a, b^+) - \operatorname{lk}(b, a^+)$ is the signed number of intersections of a with b in F, and hence the non-zero entries in M lie in blocks on the diagonal of the form

$$\begin{pmatrix} 0 & 1 \\ -1 & 0 \end{pmatrix}.$$

$\square$

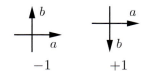

Figure 6.7. Intersection number $i(a, b)$ for curves a and b.

Theorem 6.7.7. *If S is a satellite knot constructed from pattern P with companion C, framing zero, and winding number n then*

$$\det(S) = \begin{cases} \det(P) & \text{if } n \text{ is even,} \\ \det(P)\det(C) & \text{if } n \text{ is odd.} \end{cases}$$

PROOF. Let V be the companion solid torus associated with C, and let h be the faithful homeomorphism that maps the pattern $P \subset W$ to $S \subset V$.

Let F be a surface spanning S. In the proof of Theorem 5.8.1 we showed how F can be arranged so that it meets ∂V in n parallel preferred longitudes. The part of F outside V consists of n parallel copies of a surface F_C spanning the companion. The preimage $h^{-1}(F \cap V)$ is a surface in the unknotted solid torus W that can be made into a spanning surface F_P for P by adding n parallel discs outside W.

A basis for $H_1(F_P)$ can be isotoped in F_P so that it lies in W (any loop outside W bounds a disc and so is trivial in homology). The images of these basis elements under h are some of the basis elements we want for $H_1(F)$. Let M_P be the Seifert matrix for F_P. The linking numbers between the loops in $H_1(F_P)$ are unchanged by h so there is a copy of M_P sitting inside the Seifert matrix for F. To complete the basis for F take a basis for one sheet of $F - V$ and replicate the basis in the other sheets by translation. Let M_C be the Seifert matrix for F_C.

Take loops $\lambda_P \subset V \cap F$ and λ_C in one of the sheets of F outside V. Now λ_P is homologous in V to a sum of longitudes of ∂V, and ∂F_C is also a longitude. By lifting λ_C out of the surface we see (Corollary 5.7.4) that $\mathrm{lk}(\lambda_P, \lambda_C) = 0$.

A typical Seifert matrix for F has the form

$$M_S = \begin{pmatrix} M_P & 0 \\ 0 & X \end{pmatrix} \quad \text{where} \quad X = \begin{pmatrix} M_C & M_C & M_C & M_C \\ M_C^{\mathrm{T}} & M_C & M_C & M_C \\ M_C^{\mathrm{T}} & M_C^{\mathrm{T}} & M_C & M_C \\ M_C^{\mathrm{T}} & M_C^{\mathrm{T}} & M_C^{\mathrm{T}} & M_C \end{pmatrix}.$$

The matrix X is composed of n^2 blocks, each of which is either M_C or its transpose; in the example here we have used $n = 4$. The blocks on the diagonal correspond to the individual sheets of $F - V$; the off-diagonal blocks come from interlacing between the sheets.

We now need to consider

$$X + X^{\mathrm{T}} = \begin{pmatrix} M_C + M_C^{\mathrm{T}} & 2\,M_C & 2\,M_C & 2\,M_C \\ 2\,M_C^{\mathrm{T}} & M_C + M_C^{\mathrm{T}} & 2\,M_C & 2\,M_C \\ 2\,M_C^{\mathrm{T}} & 2\,M_C^{\mathrm{T}} & M_C + M_C^{\mathrm{T}} & 2\,M_C \\ 2\,M_C^{\mathrm{T}} & 2\,M_C^{\mathrm{T}} & 2\,M_C^{\mathrm{T}} & M_C + M_C^{\mathrm{T}} \end{pmatrix}.$$

The value of the following sum of blocks is independent of the column:

$$A = \sum_{i=1}^{n} (-1)^{i+1} \,(\text{row } i) = M_C + (-1)^{n+1}\,M_C^{\mathrm{T}}.$$

Replace row 1 with this alternating sum of the rows:

$$\begin{pmatrix} A & A & A & A \\ 2\,M_C^{\mathrm{T}} & M_C + M_C^{\mathrm{T}} & 2\,M_C & 2\,M_C \\ 2\,M_C^{\mathrm{T}} & 2\,M_C^{\mathrm{T}} & M_C + M_C^{\mathrm{T}} & 2\,M_C \\ 2\,M_C^{\mathrm{T}} & 2\,M_C^{\mathrm{T}} & 2\,M_C^{\mathrm{T}} & M_C + M_C^{\mathrm{T}} \end{pmatrix}.$$

For each $i > 1$, replace column i with column i minus column 1:

$$\begin{pmatrix} M_C + (-1)^{n+1}\,M_C^{\mathrm{T}} & 0 & 0 & 0 \\ 2\,M_C^{\mathrm{T}} & M_C - M_C^{\mathrm{T}} & 2(M_C - M_C^{\mathrm{T}}) & 2(M_C - M_C^{\mathrm{T}}) \\ 2\,M_C^{\mathrm{T}} & 0 & M_C - M_C^{\mathrm{T}} & 2(M_C - M_C^{\mathrm{T}}) \\ 2\,M_C^{\mathrm{T}} & 0 & 0 & M_C - M_C^{\mathrm{T}} \end{pmatrix}.$$

The determinant is the product of the determinants of the diagonal blocks:

$$\det(M_S + M_S^{\mathrm{T}})$$
$$= \det(M_P + M_P^{\mathrm{T}})\,\det(M_C + (-1)^{n+1}M_C^{\mathrm{T}})\,\left[\det(M_C - M_C^{\mathrm{T}})\right]^{n-1}.$$

Since the companion C must be a knot, we can apply Lemma 6.7.6 to complete the proof. $\square$

$$D_+ \qquad\qquad D_- \qquad\qquad D_0$$

Figure 6.8. Three oriented diagrams which differ only inside a small neighbourhood.

6.8   Signature and unknotting number

The development of link invariants derived from spanning surfaces and Seifert matrices has largely concentrated on global features. We can also examine how small localised changes made to a link affect the Seifert matrix. In particular, we shall look at unknotting moves.

The unknotting number of a knot is one of the most difficult invariants to determine. Various techniques are known which bound the unknotting number in different ways. Each of them eats away at the problem, but there is no general mechanism for calculating the unknotting number in a given case. It is clear that if a non-trivial knot can be unknotted by making only one change then its unknotting number is 1. The problem is for higher unknotting numbers: how do you prove that a solution you have found by experiment is minimal?

If we pass one strand of a knot K through another, we can always find a diagram of K to illustrate the change in which the switch happens at a crossing. This can be indicated schematically as shown in Figure 6.8. The three diagrams are assumed to be identical outside the broken circle. The change from D_+ to D_- or vice versa is an unknotting move. In the last picture, D_0, the crossing has been eliminated in the same manner as was used to construct Seifert circles; it is included since it is helpful in the proof below.

It is clear that the Seifert matrices of the projection surfaces constructed from these three diagrams will have much in common. The similarities can be exploited to find relationships between the matrix-based invariants. As a first example, we shall show that switching a crossing in a knot diagram can change the signature by at most 2 [132].

To establish this result we need the following lemma* [128]. A σ-series

---

*A proof is reproduced in Appendix A2 of [9].

for an $n \times n$ matrix of rank r is a sequence of submatrices Δ_i for $1 \leqslant i \leqslant n$ such that

1. Δ_i has i rows and i columns,
2. Δ_i is a principal submatrix of Δ_{i+1},
3. no two consecutive Δ_i and Δ_{i+1} are both singular when $i < r$.

We also define the function

$$\text{sign}(x) = \begin{cases} -1 & \text{if } x < 0, \\ 0 & \text{if } x = 0, \\ +1 & \text{if } x > 0. \end{cases}$$

Lemma 6.8.1. *Let A be an $n \times n$ symmetric matrix. It has a σ-series $\Delta_1, \ldots, \Delta_n$. Furthermore, given any such series and putting $\Delta_0 = 1$ the signature of A can be obtained as*

$$\sigma(A) = \sum_{i=1}^{n} \text{sign}(\det(\Delta_{i-1}) \det(\Delta_i)).$$

Theorem 6.8.2. *Let K_+ and K_- be two knots that have diagrams D_+ and D_- which differ by an unknotting move as shown in Figure 6.8. Then*

$$\sigma(K_+) - \sigma(K_-) = 0 \text{ or } \pm 2.$$

PROOF. Let F_+, F_- and F_0 be the projection surfaces constructed from D_+, D_- and D_0 respectively, and let M_0 be the Seifert matrix of F_0. If F_0 is disconnected then K_+ and K_- are isotopic and have the same signature.

So assume that F_0 is connected. Let c be the crossing which is switched or eliminated. The surfaces F_+ and F_- are obtained from F_0 by adding a twisted rectangle at c. Let $(a_1, \ldots, a_n)$ be a basis for $H_1(F_0)$. Each loop a_i is also a subset of F_+ and F_-. A basis for $H_1(F_+)$ can be completed by adding a loop b which passes once through the twisted rectangle and back through the rest of the surface. The Seifert matrix M_+ for F_+ has the form

$$
\begin{array}{c|ccccc}
 & a_1^+ & \cdots & a_n^+ & b^+ \\
\hline
a_1 & & & & \nu_1 \\
\vdots & & M_0 & & \vdots \\
a_n & & & & \nu_n \\
b & \lambda_1 & \cdots & \lambda_n & \beta
\end{array}
$$

Using the same loop b in F_- gives a Seifert matrix M_- for F_- that is identical to M_+ except in the bottom-right corner: in F_- the linking number $\mathrm{lk}(b, b^+) = \beta + 1$.

For simplicity, we write $A_* = M_* + M_*^{\mathrm{T}}$ where $*$ is $+$, $-$ or 0. Because K_+ is a knot, $\det(A_+)$ is odd (Theorem 6.7.1) so A_+ is non-singular. Similarly for A_-. The link ∂F_0 has two components; hence, by Exercise 6.9.13, the matrix A_0 has rank n or $n - 1$. We define σ-series for A_0 in the two cases as follows.

1. If A_0 has maximal rank, we take $\Delta_1, \ldots, \Delta_n$ to be a σ-series for A_0. Note that $\Delta_n = A_0$.

2. If A_0 is singular then we can arrange (by changing the basis if necessary) that its submatrix formed from the first $n - 1$ rows and $n - 1$ columns has maximal rank. Let $\Delta_1, \ldots, \Delta_{n-1}$ be a σ-series for this submatrix and put $\Delta_n = A_0$. Note that Δ_{n-1} is non-singular.

In both cases, putting $\Delta_{n+1} = A_+$ gives a σ-series for A_+. Similarly for A_-. Therefore

$$
\sigma(A_+) = \mathrm{sign}(\det(A_0)\det(A_+)) + \sum_{i=1}^{n} \mathrm{sign}(\det(\Delta_{i-1})\det(\Delta_i)),
$$

$$
\sigma(A_-) = \mathrm{sign}(\det(A_0)\det(A_-)) + \sum_{i=1}^{n} \mathrm{sign}(\det(\Delta_{i-1})\det(\Delta_i)).
$$

Noting that $\mathrm{sign}(xy) = \mathrm{sign}(x)\,\mathrm{sign}(y)$, we deduce that

$$
\sigma(A_+) - \sigma(A_-) = \mathrm{sign}(\det(A_0))\Big(\mathrm{sign}(\det(A_+)) - \mathrm{sign}(\det(A_-))\Big).
$$

Now $\mathrm{sign}(\det(A_0))$ can be 0 or ± 1. The matrices A_+ and A_- are both non-singular so the other signs are non-zero and their difference is even — that is 0 or ± 2.   □

Corollary 6.8.3. *If K is a knot then $u(K) \geqslant \frac{1}{2}|\sigma(K)|$.*

PROOF. An unknotting operation can change the signature by at most 2. For the trivial knot $u(\bigcirc) = \sigma(\bigcirc) = 0$.   □

Corollary 6.8.4. *Unknotting number can be arbitrarily large.*

PROOF. Let K be the $(p, 2)$ torus knot with $p \geqslant 3$. Then $\sigma(K) = p - 1$ (Exercise 6.9.8). The standard minimal-crossing diagram of K can be unknotted by switching $\frac{1}{2}(p - 1)$ crossings. Hence $u(K) = \frac{1}{2}(p - 1)$. □

If we can find by experiment a sequence of n unknotting moves which unknot K and if $n = \frac{1}{2}|\sigma(K)|$ then we know that the solution is minimal. For example, the knot 10_8 shown in Figure 3.15 has signature 4 and hence has unknotting number 2. This simple method suffices to give the unknotting numbers of 39 of the prime knots up to nine crossings. Removing the further 17 cases where the unknotting number is 1, we are left with 28 cases in which an experimental solution cannot be proved minimal using the signature or non-triviality. The first few of these are listed below; $u_*(K)$ denotes the experimentally determined upper bound on unknotting number.

	7_4	8_3	8_4	8_6	8_8	8_{10}	8_{12}	8_{16}	8_{18}
$\frac{1}{2}|\sigma(K)|$	1	0	1	1	0	1	0	1	0
$u_*(K)$	2	2	2	2	2	2	2	2	2

The first problem case is 7_4. Trial and error shows that it can be unknotted using two switches, but its signature gives a lower bound of 1. In fact, more sophisticated techniques [249] can be used to show that this knot cannot have unknotting number 1. The same is true of the other knots in this list. However, there are cases where the problem is still open: for example it is unknown whether $u(9_{10})$ is 2 or 3.

6.9 Exercises

1. Let X be the set of four 1-cycles that bound the four small triangles in the graph, G, shown below.

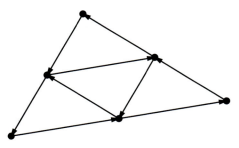

(a) Show that X is a basis for $Z_1(G)$.

(b) Show that this basis cannot be produced using the spanning-tree construction of Theorem 6.1.2.

(c) Use the spanning-tree construction to find another set of four loops, Y, that is also a basis for $Z_1(G)$.

(d) Write down the change-of-basis matrix that expresses the loops in Y as linear combinations of those in X.

2. Complete the proof of Theorem 6.1.2.

3. Show that $\partial_1 \circ \partial_2 = 0$ for all 2-chains.

4. Find a Seifert matrix for a trefoil. Compute $\det(3_1)$ and $\sigma(3_1)$.

5. Using the values in Figure 6.6 and the results of this chapter, show that the determinant and signature can distinguish most of the knots in Figure 1.6. Which three cannot be separated?

6. Let K be the (p, q, r) pretzel knot with p, q and r all odd. Find a Seifert matrix for K. Compute $\det(K)$ and $\sigma(K)$.

7. Show that these invariants cannot distinguish $P(-3, 5, 7)$ from the trivial knot. For what other values of p, q and r does this happen? Find a relationship between the values.

8. Show that the signature of the $(p, 2)$ torus knot with $p \geqslant 3$ is $p - 1$.

9. Complete the proof of Lemma 6.7.6.

10. Construct an example to show that the determinant of a Seifert matrix is not a link invariant.

11. Diagonalise the following matrix and show that its signature is zero:

$$\begin{pmatrix} 0 & 1 \\ 1 & 0 \end{pmatrix}.$$

12. Show that the linking form Θ (defined on page 139) is bilinear. Note that Θ is not symmetric.

13. The nullity of a diagonal matrix is the number of zeroes on the diagonal. Define the nullity of a link, $n(L)$, as the nullity of a diagonalised form of $M + M^{\mathrm{T}}$ where M is any Seifert matrix for L. Show that

(a) $n(L)$ is well-defined,

(b) $n(K) = 0$ for a knot K,

(c) $n(L) \leqslant \mu(L) - 1$.

14. Prove the converse of Lemma 6.7.6: if M is a matrix such that $\det(M - M^{\mathrm{T}}) = 1$ then there is a surface F with one boundary component that has M as its Seifert matrix. Start with the surface in Figure 6.3, add twists to the bands to get the entries on the diagonal of M, then interlace them appropriately to produce the other entries.

15. The arrangement of occupied and vacant cells in the table in Figure 6.6 is suggestive of a chessboard. Show that $\mathrm{sign}(\det(K)) = (-1)^{\sigma(K)/2}$ when K is a knot.

Hence deduce that $\sigma(K) \equiv \det(K) - 1 \bmod 4$ to verify the chessboard pattern.

7 The Alexander–Conway Polynomial

So far, the invariants we have seen have all been numerical — either geometrically motivated or resulting from a computation. In this chapter we shall introduce polynomial-valued invariants. First we shall derive the classical Alexander polynomial from the Seifert matrix. The Conway polynomial has an axiomatic definition based on relationships between diagrams. In spite of their different origins, a change of variable shows that these two polynomials are actually only one invariant. The Alexander viewpoint allows us to obtain a lower bound on the genus of a link, while Conway's algebra of tangles explains some of the symmetry properties of the polynomial.

7.1 The Alexander polynomial

Definition 7.1.1. The *Alexander polynomial*[*] of an oriented link L, written variously as $\Delta(L)$ or $\Delta_L(x)$ depending on the required emphasis, is the determinant $\det(xM - x^{-1}M^{\mathrm{T}})$ where M is any Seifert matrix for L. It is a Laurent polynomial in $\mathbb{Z}[x^{\pm 1}]$, that is a polynomial in positive and negative powers of x with integer coefficients.

As with the determinant and signature definitions, it is not obvious that the Alexander polynomial of a link is independent of M. The proof that it is well-defined is essentially the same as the proof of Theorem 6.6.5.

Theorem 7.1.2. *Let M be a Seifert matrix constructed from a surface F spanning an oriented link L. Then $\det(xM - x^{-1}M^T)$ is a link invariant.*

---

[*]The Alexander polynomial is usually defined as $\Delta(t) = \det(t^{1/2}M - t^{-1/2}M^{\mathrm{T}})$ so that it agrees with the value obtained using a different construction. This is related to the version used here by $\Delta(x) = \Delta(t^2)$. Using our form we get rid of the square roots. The version defined here is also a normalised form; when defined in other ways the polynomial is only determined up to multiplication by powers of $\pm t$.

PROOF. To prove that $\Delta(L)$ is independent of the choice of F and the basis for $H_1(F)$ that gives M, we need to check invariance under congruence and enlargement operations. Congruence is straightforward:

$$
\begin{aligned}
\det(x\,P^{\mathrm{T}}MP - x^{-1}(P^{\mathrm{T}}MP)^{\mathrm{T}}) &= \det(P^{\mathrm{T}}(xM - x^{-1}M^{\mathrm{T}})P) \\
&= \det(P^{\mathrm{T}})\det(tM - x^{-1}M^{\mathrm{T}})\det(P) \\
&= \det(xM - x^{-1}M^{\mathrm{T}}).
\end{aligned}
$$

Applying Lemma 6.6.1 we see that an enlarged matrix has the matrix $xM - x^{-1}M^{\mathrm{T}}$ in the top-left corner, and the following matrix (or its transpose) in the bottom-right:

$$
\begin{pmatrix} 0 & x \\ -x^{-1} & 0 \end{pmatrix}.
$$

Because the only non-zero entries in the last row and last column are in the bottom-right block, expanding the determinant gives $\det(xM - x^{-1}M^{\mathrm{T}})$ as desired.   □

Even though there are infinitely many knots with a given polynomial (Corollary 7.4.4), in practice, the Alexander polynomial is a useful tool for distinguishing links. Of the knots in Figure 6.6, only five pairs remain unresolved:

$$
\begin{aligned}
\Delta(6_1) &= \Delta(9_{46}) = -2\,x^2 + 5 - 2\,x^{-2}, \\
\Delta(7_4) &= \Delta(9_2) = 4\,x^2 - 7 + 4\,x^{-2}, \\
\Delta(8_{14}) &= \Delta(9_8) = -2\,x^4 + 8\,x^2 - 11 + 8\,x^{-2} - 2\,x^{-4}, \\
\Delta(8_{18}) &= \Delta(9_{24}) = -x^6 + 5\,x^4 - 10\,x^2 + 13 - 10\,x^{-2} + 5\,x^{-4} - x^{-6}, \\
\Delta(9_{28}) &= \Delta(9_{29}) = x^6 - 5\,x^4 + 12\,x^2 - 15 + 12\,x^{-2} - 5\,x^{-4} + x^{-6}.
\end{aligned}
$$

Looking at these examples, we see that only even powers of x appear, and that the coefficients are symmetric. These are general properties of the Alexander polynomial.

Theorem 7.1.3.   *If L is a link with μ components then*

(a) $\Delta_L(-x) = (-1)^{\mu-1}\,\Delta_L(x)$,

(b) $\Delta_L(x^{-1}) = (-1)^{\mu-1}\,\Delta_L(x)$.

PROOF. These relations are consequences of the following properties of the matrix determinant: if A is a square matrix then $\det(A) = \det(A^{\mathrm{T}})$ and $\det(-A) = (-1)^n \det(A)$ where n is the number of rows in A.

If a surface F spanning L has genus g then the Seifert matrix M for F has $2g + \mu - 1$ rows.

Let $A = xM - x^{-1}M^{\mathrm{T}}$. Part (a) is the relation between $\det(A)$ and $\det(-A)$; part (b) is the relation between $\det(A)$ and $\det(-A^{\mathrm{T}})$. $\qquad\square$

Hence, for a knot K, the Alexander polynomial is an even function of x and only powers of x^2 appear. Also $\Delta_K(x^{-1}) = \Delta_K(x)$ so the coefficients are symmetric.

All the properties of the determinant established in §6.7 follow through in a straightforward manner for the Alexander polynomial. These are collected in the following theorem.

Theorem 7.1.4. *The Alexander polynomial has the following properties.*

(a) *If L is a link then $|\Delta_L(\pm i)| = \det(L)$ where $i = \sqrt{-1}$.*

(b) *If K is a knot then $\Delta_K(1) = 1$.*

(c) *If L is a link with more than one component then $\Delta_L(1) = 0$.*

(d) *If L is a link then $\Delta(-L) = \Delta(L) = \Delta(L^*)$.*

(e) *If $L_1 \sqcup L_2$ is a split link then $\Delta(L_1 \sqcup L_2) = 0$.*

(f) *If a link can be factorised as $L_1 \# L_2$ then $\Delta(L_1 \# L_2) = \Delta(L_1)\,\Delta(L_2)$.*

(g) *If S is a satellite knot constructed from pattern P with companion C, framing zero, and winding number n then $\Delta_S(x) = \Delta_P(x)\,\Delta_C(x^n)$.*

PROOF. Part (a) follows immediately from the definitions. Parts (b) and (c) are corollaries of Lemma 6.7.6.

The proofs of (d), (e) and (f) are essentially the same as the proofs of Theorems 6.7.2, 6.7.4 and 6.7.5 respectively.

For part (g), the proof is modelled on that of Theorem 6.7.7 and uses the same notation. The Seifert matrix of the satellite is the same and has the form

$$\begin{pmatrix} M_P & 0 \\ 0 & X \end{pmatrix} \qquad \text{where} \qquad X = \begin{pmatrix} M_C & M_C & M_C & M_C \\ M_C^{\mathrm{T}} & M_C & M_C & M_C \\ M_C^{\mathrm{T}} & M_C^{\mathrm{T}} & M_C & M_C \\ M_C^{\mathrm{T}} & M_C^{\mathrm{T}} & M_C^{\mathrm{T}} & M_C \end{pmatrix}.$$

As before, the matrix X is composed of $n \times n$ blocks, each of which is either M_C or its transpose. We shall continue to use $n = 4$ as the example case. If g is the genus of the surface F_C spanning the companion, then M_C has $2g$ rows and columns.

We now need to consider the matrix $x X - x^{-1} X^{\mathrm{T}}$:

$$
\begin{pmatrix}
x\, M_C - x^{-1} M_C^{\mathrm{T}} & x\, M_C - x^{-1} M_C & x\, M_C - x^{-1} M_C & x\, M_C - x^{-1} M_C \\
x\, M_C^{\mathrm{T}} - x^{-1} M_C^{\mathrm{T}} & x\, M_C - x^{-1} M_C^{\mathrm{T}} & x\, M_C - x^{-1} M_C & x\, M_C - x^{-1} M_C \\
x\, M_C^{\mathrm{T}} - x^{-1} M_C^{\mathrm{T}} & x\, M_C^{\mathrm{T}} - x^{-1} M_C^{\mathrm{T}} & x\, M_C - x^{-1} M_C^{\mathrm{T}} & x\, M_C - x^{-1} M_C \\
x\, M_C^{\mathrm{T}} - x^{-1} M_C^{\mathrm{T}} & x\, M_C^{\mathrm{T}} - x^{-1} M_C^{\mathrm{T}} & x\, M_C^{\mathrm{T}} - x^{-1} M_C^{\mathrm{T}} & x\, M_C - x^{-1} M_C^{\mathrm{T}}
\end{pmatrix}.
$$

The following summation ranges over blocks in a column of this matrix; its value is independent of the column chosen:

$$
\sum_{i=1}^{n} x^{n+1-2i}\, (\text{row } i) \;=\; x^n\, M_C - x^{-n}\, M_C^{\mathrm{T}}.
$$

Replace row 1 with this sum of the rows, and divide by $x^{2g(n-1)}$ to preserve the determinant. Now replace column i with column i minus column 1, for each $i > 1$. This produces the matrix

$$
x^{-2g(n-1)}
\begin{pmatrix}
x^n\, M_C - x^{-n}\, M_C^{\mathrm{T}} & 0 & 0 & 0 \\
* & x(M_C - M_C^{\mathrm{T}}) & * & * \\
* & 0 & x(M_C - M_C^{\mathrm{T}}) & * \\
* & 0 & 0 & x(M_C - M_C^{\mathrm{T}})
\end{pmatrix}.
$$

The asterisks indicate values which do not affect the determinant. Since C is a knot, Lemma 6.7.6 shows that in $n - 1$ blocks on the diagonal we have $\det(x(M_C - M_C^{\mathrm{T}})) = x^{2g}$. Hence

$$
\det(x X - x^{-1} X^{\mathrm{T}}) \;=\; \det(x^n\, M_C - x^{-n}\, M_C^{\mathrm{T}}).
$$

So the determinant is the product of the determinants of the diagonal blocks:

$$
\det(x\, M_S - x^{-1} M_S^{\mathrm{T}}) \;=\; \det(x\, M_P - x^{-1} M_P^{\mathrm{T}})\, \det(x^n\, M_C - x^{-n} M_C^{\mathrm{T}}).
$$

$\square$

7.2 The Alexander polynomial and genus

The definition and calculation of the Alexander polynomial rely on spanning surfaces so it should not be much of a surprise to discover that there is a relationship between the genus of a link and its Alexander polynomial. In fact, in many cases, the following result can be used to prove that a given spanning surface actually has minimal genus.

The *breadth* of a polynomial is the difference between its highest and lowest degrees.

Theorem 7.2.1. *The genus of a non-split link L is bounded below by the breadth of the Alexander polynomial:*

$$2g(L) + \mu(L) - 1 \geqslant \tfrac{1}{2} \text{ breadth } \Delta_L(x).$$

PROOF. Let F be a connected minimal-genus spanning surface for L and let $r = 2g + \mu - 1$. A basis for $H_1(F)$ has r generators, so the Seifert matrix for F is an $r \times r$ square matrix. Thus the largest possible degree for x is r and the smallest degree is $-r$. Hence breadth $\Delta_L(x) \leqslant 2r$.

Now

$$
\begin{aligned}
2\,g(F) &= 2 - \chi(F) - \mu(L) \\
&= 2r - \mu(L) + 1 \\
&\geqslant \tfrac{1}{2} \text{ breadth } \Delta_L(x) - \mu(L) + 1.
\end{aligned}
$$

$\square$

Consider again the knot 10_{165} shown in Figure 6.4.

$$\Delta(10_{165}) = 3x^4 - 11x^2 + 17 - 11x^{-2} + 3x^{-4}.$$

The projection surface in the figure has genus 3, but the bound we deduce from the Alexander polynomial is 2. In fact, 10_{165} does have genus 2: loop b bounds a compressing disc for the projection surface.[*]

In fact, this theorem is powerful enough to establish the genus for all knots up to 10 crossings. However, there are 11-crossing knots for which it fails. For example, if K is the Kinoshita–Terasaka knot shown in Figure 4.12 then $c(K) = 11$ and $\Delta(K) = 1$. This knot is non-trivial (see page 234) so its genus is at least 1 (in fact $g(K) = 2$ [142]).

---

[*]We already remarked that a minimal-genus projection surface for 10_{165} cannot be constructed from a minimal-crossing diagram — see page 105.

7.3   The Alexander skein relation

In §6.8 we saw how the Seifert matrices of two similar knots, K_+ and K_-, are related. We can try this approach again with the Alexander polynomial. This time we do not get information on the unknotting number; the result has far greater consequences and provides a whole new approach to defining and calculating polynomial invariants.

Figure 6.8 shows small parts of three link diagrams which are identical outside a small neighbourhood. The broken circle indicates the boundary of the neighbourhood, and the links differ inside as shown. The diagrams can be transformed into one another by replacing one picture with another. A change from D_+ to D_- or vice versa is called *switching* a crossing. A change from D_+ or D_- to D_0 is called *smoothing* a crossing. Note that these local substitution operations can (and probably do) change the link type. Smoothing a crossing always increases or decreases the number of components by 1.

Theorem 7.3.1.   *If three oriented links L_+, L_- and L_0 have diagrams D_+, D_- and D_0 which differ only in a small neighbourhood as shown in Figure 6.8 then*

$$\Delta(L_+) - \Delta(L_-) \;=\; (x^{-1} - x)\,\Delta(L_0).$$

PROOF. Let F_+, F_- and F_0 be the projection surfaces constructed from D_+, D_- and D_0 respectively, and let M_0 be the Seifert matrix of F_0.

If D_+ is a disconnected diagram then D_- and D_0 must also be disconnected. This implies that L_+, L_- and L_0 are all split links. So, from Theorem 7.1.4(e), the relation holds. The same argument applies when D_- is a disconnected diagram.

Suppose that D_0 is a disconnected diagram but that D_+ and D_- are connected. Then L_0 is a split link and the links L_+ and L_- are isotopic — simply turn over the right-hand part of the diagram in Figure 7.1. Again the relation holds.

In the remaining case, all the diagrams D_+, D_- and D_0 are connected. Let $(a_1, \ldots, a_n)$ be a basis for $H_1(F_0)$. Each loop a_i is also a subset of F_+ and F_-. A basis for $H_1(F_+)$ can be completed by adding a loop b which passes once through the twisted band in the tangle and back through the

Figure 7.1.

rest of the surface. The Seifert matrix M_+ for F_+ has the form

$$
\begin{array}{c|ccccc}
 & a_1^+ & \cdots & a_n^+ & b^+ \\
\hline
a_1 & & & & \nu_1 \\
\vdots & & M_0 & & \vdots \\
a_n & & & & \nu_n \\
b & \lambda_1 & \cdots & \lambda_n & \beta
\end{array}
$$

Using the same loop b in F_- gives a Seifert matrix M_- for F_- that is identical to M_+ except in the bottom-right corner: in F_- the linking number $\mathrm{lk}(b, b^+) = \beta + 1$.

Expanding the determinants $\det(xM_+ - x^{-1}M_+^{\mathrm{T}})$ and $\det(xM_- - x^{-1}M_-^{\mathrm{T}})$ about the last column and subtracting terms, we see that almost everything cancels and we are left with

$$
\begin{aligned}
&\Delta(L_+) - \Delta(L_-) \\
={}& \beta(x - x^{-1}) \det(xM_0 - x^{-1}M_0^{\mathrm{T}}) \\
&- (\beta + 1)(x - x^{-1}) \det(xM_0 - x^{-1}M_0^{\mathrm{T}}) \\
={}& -(x - x^{-1}) \det(xM_0 - x^{-1}M_0^{\mathrm{T}}) \\
={}& (x^{-1} - x) \Delta(L_0).
\end{aligned}
$$

$\square$

This relationship between three links with local differences is an example of a *skein relation*. Note that the relationship depends only on the link types, and not on the particular diagrams.

7.4   The Conway polynomial

The skein relation can be used as the basis of an axiomatic definition.

Definition 7.4.1.   The *Conway polynomial* of an oriented link L, denoted variously by $\nabla(L)$ or $\nabla_L(z)$ depending on the required emphasis, is defined by the three following axioms.

1. Invariance: $\nabla_L(z)$ is invariant under ambient isotopy of L.

2. Normalisation: if K is the trivial knot then $\nabla_K(z) = 1$.

3. Skein relation: $\nabla(L_+) - \nabla(L_-) = z\,\nabla(L_0)$ where L_+, L_- and L_0 have diagrams D_+, D_- and D_0 which differ as shown in Figure 6.8.

It is a polynomial in $\mathbb{Z}[z]$.

These axioms suffice for a recursive computation of $\nabla_L(z)$. The skein relation can be rewritten in two forms:

$$\begin{aligned}
\nabla(L_+) &= \nabla(L_-) + z\,\nabla(L_0), \\
\nabla(L_-) &= \nabla(L_+) - z\,\nabla(L_0).
\end{aligned}$$

By switching crossings, any link can be transformed into a trivial link (Exercise 3.10.4). Smoothing reduces the number of crossings and so, inductively, the process proceeds to simpler links. By repeatedly applying the skein relation, we can ultimately express the Conway polynomial of a given link in terms of trivial links.

As an example, let us calculate $\nabla(4_1)$. The calculation is shown in Figure 7.2. At the top of the tree is a diagram of the knot whose polynomial we want to calculate. The skein relation (axiom 3) is applied to the positive crossing at the top-right: the results of switching and smoothing the crossing are shown on the next level of the tree. Invariance under ambient isotopy (axiom 1) means that we can simplify these diagrams without affecting their polynomials. The diagram on the left (produced by switching the crossing) is clearly ambient isotopic to the trivial knot. By axiom 2 this has $\nabla(z) = 1$ so this side of the tree can stop. The diagram on the right (produced by smoothing the crossing) is a Hopf link and can be simplified as shown. The polynomial of this is unknown so we need to apply the skein relation again. This time a negative crossing is used and the products are a 2-component trivial link and another trivial knot. All the terminal nodes of the tree are now labelled with trivial links. The next result shows that $\nabla(z) = 0$ on a split link. Each right-hand branch of the tree is labelled with $\pm z$ according to whether a positive or a negative crossing was smoothed. The polynomial is constructed by taking the polynomials of the links at the terminal nodes and combining them. In this case we see that $\nabla(4_1) = 1 + z\,(0 - z \cdot 1) = 1 - z^2$.

Theorem 7.4.2. *If L is a split link then $\nabla(L) = 0$.*

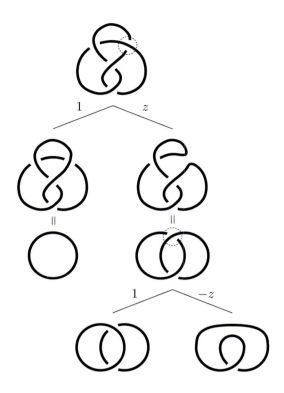

Figure 7.2. A resolving tree for the Conway polynomial of the figure-8 knot.

PROOF. Let D_0 be a disconnected diagram of the split link $L_1 \sqcup L_2$ arranged so that projections of L_1 and L_2 are disjoint, and so that the neighbourhood defining D_0 contains one arc from each L_i. Then D_+ will be arranged as shown in Figure 7.1, and D_- is obtained by turning the right half of the diagram through a full twist. Therefore D_+ and D_- are isotopic and, by axiom 1, $\nabla(D_+) = \nabla(D_-)$. Applying axiom 3 completes the argument:

$$z \, \nabla(D_0) \;=\; \nabla(D_+) - \nabla(D_-) \;=\; 0.$$

$\square$

The axioms must be consistent to be useful — it is no good if two calculations made on the same link can give rise to different results. The consistency can be verified in a purely combinatorial manner using induction on the number of crossings, establishing the invariance of $\nabla(L)$ under Reidemeister moves, order of evaluation, and so on [193]. However, making

Figure 7.3.

the substitution $z = x^{-1} - x$, we recover the Alexander polynomial, and all the axioms are known to hold in this case.

If K is a knot then $\nabla_K(z)$ is a polynomial in z^2. It is a genuine polynomial (only positive powers appear), and the constant term is always 1. Within these constraints, all possible polynomials can be realised as the Conway polynomial of some knot.

Proposition 7.4.3. *The knot K shown in Figure 7.3 has Conway polynomial*

$$\nabla_K(z) = 1 + (-1)^{n+1} a_n z^{2n} + \sum_{i=1}^{n-1} (-1)^{i+1} (a_i - 1) z^{2i}.$$

PROOF. The proof proceeds by induction on n (Exercise 7.12.8). Concentrate on the left-hand end of the knot. First reduce the number of twists in box a_1: switching a crossing reduces the number of twists, smoothing a crossing produces a Hopf link. The proof needs to cope with the cases of a_i positive, negative and zero. ☐

Note that this produces only one example for each polynomial: even though all sequences of coefficients a_i of the form $(1, \dots, 1, 0)$ give $\nabla_K(z) = 1$, the corresponding diagrams all depict the trivial knot. Similarly, appending $(1, \dots, 1, 0)$ to any sequence does not change the polynomial or the knot.

Corollary 7.4.4. *There are infinitely many knots with a given Conway polynomial.*

PROOF. Let K_A be the knot shown in Figure 7.3 with the given polynomial, and let K_B be the untwisted double of the figure-8 knot. Our examples will be formed by taking the products

$$K_A, \ K_A \# K_B, \ K_A \# K_B \# K_B, \ K_A \# K_B \# K_B \# K_B, \ \dots$$

These knots all have the same Conway polynomial since $\nabla(K_B) = 1$ (Exercise 7.12.1) and, by Theorem 7.1.4(e), $\nabla(K_A \# K_B) = \nabla(K_A) \nabla(K_B)$.

To show that the knots are distinct, we show they have different genera. The figure-8 knot is shown to be non-trivial by its determinant. Theorem 4.4.1 shows that the double of a non-trivial knot is non-trivial. A non-trivial knot has genus at least 1, and a double knot has a genus-1 spanning surface, hence $g(K_B) = 1$. Applying Theorem 5.6.1 we see that the genus of the product of K_A with n copies of K_B is $g(K_A) + n$. □

The knots constructed in the preceding corollary are neither prime nor simple. It is possible to find infinitely many simple knots with a given Conway polynomial [141] but showing that the knots are, in fact, distinct is then much more difficult.

Corollary 7.4.5. *There is a knot with unknotting number* 1 *and any given Conway polynomial.*

PROOF. The Kinoshita–Terasaka knot shown in Figure 4.12 has Conway polynomial $\nabla(z) = 1$ and unknotting number 1 (switch the central crossing). For other polynomials, we use the knot in Figure 7.3: switching one of the two left-most crossings produces the trivial knot. □

Corollary 7.4.6. *There are infinitely many knots with a given genus.*

PROOF. When $a_n \neq 0$, the knot shown in Figure 7.3 has genus n (Exercise 7.12.9). □

Corollary 7.4.7. *Knots with unknotting number* 1 *can have arbitrarily high genus.*

These examples show that the unknotting number is not directly related to the genus or the Conway polynomial. It is common for knot theorists to construct examples that demonstrate the independence of invariants in this manner.

7.5 Resolving trees

The process of evaluating a knot polynomial by repeated application of a skein relation is called *resolution*. The knot is resolved into a linear combination of polynomials and trivial links. The process can be easily recorded schematically in a binary tree structure as shown in Figure 7.2. The link we start with is placed at the root of the tree, each application of the skein relation is recorded as a bifurcation so that each triplet (parent, left child, right child) corresponds to either (D_+, D_-, D_0) or (D_-, D_+, D_0). The Conway polynomial of the root link is then expressed as a combination of the links at the terminal nodes.

Resolution may seem to be a fairly arbitrary affair in which a good choice of crossing can save lots of calculation. However, it is possible to be systematic about it [140].

Lemma 7.5.1. *It is possible to construct a resolving tree for a link so that in any path from the root to a terminal node, no crossing is changed more than once.*

PROOF. Let $L = K_1 \cup \cdots \cup K_\mu$ be a link with μ components. Let D be a diagram of L and let D_i be subdiagrams of D so that each D_i is a diagram of L_i.

Orient the link and choose a basepoint on each D_i distinct from any crossing in D. For each i in sequence, follow the diagram in the direction of its orientation until a crossing c is reached which is first encountered as an over-crossing. Let $A(D)$ denote the subset of D traversed before reaching c. This is the ascending set. Let $N(D) = D - A(D)$ denote the non-ascending set which contains both the over- and the under-crossing strands at c. Let $|A(D)|$ denote the number of crossings of D through which $A(D)$ passes (for each crossing counted, the under-crossing strand will be in $A(D)$ and the over-crossing strand may be), and let $|N(D)|$ denote the number of crossings where neither strand is in $A(D)$. Then $|A(D)| + |N(D)| = c(D)$.

The result is proved by induction on $|N(D)|$.

If $|N(D)| = 0$ then D is an ascending diagram and hence L is a trivial link. If $|N(D)| \neq 0$ then $N(D)$ contains a crossing c that separates D into $A(D)$ and $N(D)$. Let $D_\pm$ be the diagram obtained from D by switching the sense of c. The orientation, basepoints, and ordering on the components of $D_\pm$ can be taken to be the same as those for D since the underlying projections of D and $D_\pm$ are identical. Now $A(D_\pm) \supset A(D)$ so $|A(D_\pm)| > |A(D)|$ and $c(D_\pm) = c(D)$. So

$$|N(D_\pm)| \;=\; c(D_\pm) - |A(D_\pm)| \;<\; c(D) - |A(D)| \;=\; |N(D)|.$$

Let D_0 be the diagram obtained from D by smoothing the crossing c. The orientation of D_0 is induced from that of D. The number of components in

D_0 is one more or one less than in D, depending on whether the two strands at c are from the same subdiagram of D.

To put an ordering on the subdiagrams of D_0, give the unaltered subdiagrams the same index in the ordering and the same basepoint as they have in D. For the altered components we consider two cases.

1. Both strands at c are from the same subdiagram.

   Suppose D_r is the subdiagram that has been disconnected. One component will contain the basepoint of D_r: give this component rth place in the ordering, and give it the same basepoint as D_r. Place the other component $(\mu + 1)$th in the ordering, add a basepoint on it where c has been removed, and orient it coherently with D.

2. The strands at c are from different subdiagrams.

   Let D_r and D_s be the subdiagrams that have been joined and suppose that $r < s$. Place this new subdiagram rth in the ordering and give it the same basepoint as D_r. (Note the ordered set of subdiagrams now has no element of index s.)

Now $A(D_0) \supseteq A(D)$ so $|A(D_\pm)| \geqslant |A(D)|$ and $c(D_0) = c(D) - 1$. So

$$|N(D_0)| = c(D_0) - |A(D_0)| < c(D) - |A(D)| = |N(D)|.$$

The diagram D has been partially resolved into two diagrams $D_\pm$ and D_0 with both $|N(D_\pm)|$ and $|N(D_0)|$ less than $|N(D)|$. This has been performed without switching any crossing in $A(D)$, and so that both $A(D_\pm)$ and $A(D_0)$ contain $A(D)$. This step can be repeated inductively to give the required resolution. $\square$

Say that a polynomial is *positive* if all its non-zero coefficients are positive.

Corollary 7.5.2. *A positive link has a positive Conway polynomial.*

Proof. Take a positive diagram for the link and construct a resolving tree using the method described in the lemma. Each application of the skein relation is of the form

$$\nabla(L_+) = \nabla(L_-) + z \nabla(L_0).$$

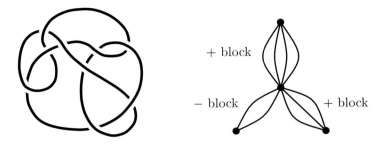

Figure 7.4. The knot 9_{43} and its Seifert graph of three blocks. All the edges in each block have the same sign, hence the diagram is homogeneous.

At every terminal node there will be a trivial knot or a split link with polynomial 1 or 0 respectively. Therefore the calculation of $\nabla(z)$ is a sum of positive terms. □

This means, for example, that no diagram of the figure-8 knot is positive.

7.6   Homogeneous links

Since the definition of the Conway polynomial is diagram-based, it should be possible to relate it to other diagram-based constructions such as Seifert's algorithm. In this section we shall explore this and show that, for a large class of links, a projection surface does have minimal genus.

Definition 7.6.1 (homogeneous link).   Let D be a diagram of an oriented link L and let G be its Seifert graph. Each edge in G can be given a sign $+$ or $-$ depending on whether it passed through a positive or negative crossing in D. If the signs of all the edges in each block of G are the same, we say that D is a homogeneous diagram [140]. A link is homogeneous if it has a homogeneous diagram.

Clearly, positive links are homogeneous because all the edges in G will be labelled $+$. Alternating diagrams are also homogeneous (Exercise 7.12.11). Figure 7.4 shows that the knot 9_{43} is homogeneous. Its Conway polynomial is $\nabla(z) = 1 + z^2 - 3z^4 - z^6$ so, by Corollary 7.5.2, 9_{43} does not have a positive diagram. We shall see at the end of §9.6 that 9_{43} is not alternating.

Theorem 7.6.2.   *Let D be a connected homogeneous diagram of a link L. Let F be the projection surface constructed from D and let G be its Seifert* graph. Then the Conway polynomial $\nabla_L(z)$ has a term of degree rank(G) in z, and it is the term of highest degree.

Furthermore, suppose that G is the union of m blocks: $G = B_1 \cup \cdots \cup B_m$. Let $\varepsilon(B_i)$ denote the sign (± 1) of the edges in block B_i. Then the highest-degree term has sign

$$\prod_{i=1}^{m} \varepsilon(B_i)^{\mathrm{rank}(B_i)}.$$

Before proving this theorem, let us see its consequences.

Corollary 7.6.3. *A homogeneous diagram is a diagram of the trivial knot if and only if its Seifert graph is a tree.*

PROOF. Let D be a diagram of the knot K with Seifert graph G. If D is homogeneous then $\nabla_K(z)$ has a term of degree rank(G) in z. If K is the trivial knot then $\nabla_K(z) = 1$ so rank$(G) = 0$ and G is a tree.

Conversely, if G is a tree then D is a diagram of the trivial knot (Exercise 5.9.14). □

This allows us to prove that many knots are non-trivial by a quick test on a diagram. For example, all the prime knots up to nine crossings except 8_{20}, 8_{21}, 9_{42}, 9_{44}, 9_{45}, 9_{46} and 9_{48} have homogeneous diagrams [140] and hence can be shown to be non-trivial 'by inspection'.

Corollary 7.6.4. *A homogeneous diagram is a diagram of a split link if and only if the diagram is disconnected.*

PROOF. Let D be a diagram of the link L with Seifert graph G. Clearly, if D is disconnected then it is a diagram of a split link.

Conversely, suppose D is a connected homogeneous diagram. Then G is also connected. Furthermore, as L is not the trivial knot (it has more than one component), G cannot be a tree and so rank$(G) > 0$. Applying the theorem shows that $\nabla_K(z)$ has a non-zero term of degree rank(G) and hence is not the polynomial of a split link (Theorem 7.4.2). □

This shows that the Whitehead link and the Borromean rings are non-trivial.

PROOF OF THEOREM 7.6.2. We prove the theorem by induction on the number of crossings in D. When $c(D) = 0$ the Seifert graph consists of a single vertex and L is the trivial knot. In this case $\nabla(z) = 1$ and $\mathrm{rank}(G) = 0$ so the theorem holds.

Now suppose that $c(D) > 0$. Choose a basepoint on each component of L in the diagram D so that D is not an ascending diagram with respect to these basepoints. Construct a resolving tree using the method described in Lemma 7.5.1. This means that no crossing is switched more than once on any path from the root of the tree to one of its leaves. The diagrams of the trivial links associated to all but the left-most leaf have fewer crossings than D.

What happens to the Seifert graph when we switch or smooth a crossing? Let G_+, G_- and G_0 be the Seifert graphs constructed from diagrams D_+, D_- and D_0. The underlying (unsigned) graphs of G_+ and G_- are equal — the graphs differ only in the sign of a single edge. When a crossing is smoothed, the corresponding edge is deleted. Hence G_0 has one less edge than G_+. Furthermore, if the crossing is smoothed in a homogeneous diagram then D_0 is also a homogeneous diagram.

To obtain terms involving powers of z, crossings must be smoothed in D or, equivalently, edges must be deleted from G. To obtain a term of as high degree as possible we must delete as many edges as possible. However, if more than $\mathrm{rank}(G)$ edges are deleted from G then the graph becomes disconnected and corresponds to a split link, for which $\nabla(z) = 0$. Similarly, any other diagram whose Seifert graph is disconnected cannot contribute to $\nabla_L(z)$. Hence any terms of degree $\mathrm{rank}(G)$ in z come from leaves in the resolving tree where the Seifert graph is a tree.

If a Seifert graph is a tree, the corresponding diagram is a diagram of the trivial knot, whose Conway polynomial is 1. When G is reduced to a tree, each block B_i is also reduced to a tree. This requires the deletion of $\mathrm{rank}(B_i)$ edges. Hence all terms of degree $\mathrm{rank}(G)$ in z at the terminal nodes are equal and have coefficient

$$\prod_{i=1}^{m} \varepsilon(B_i)^{\mathrm{rank}(B_i)}.$$

Thus they do not cancel.

It remains to show that there is at least one such contribution to $\nabla(z)$. We shall concentrate on the right-most branch of the resolving tree. At the end of this branch is a homogeneous diagram of a trivial link with fewer crossings than D. Therefore, by induction, we can apply Corollaries 7.6.3 and 7.6.4 to it.

Suppose that the trivial link at the terminal node has two or more components. Then the Seifert graph is disconnected. Following the path in the resolution down from the root, we find the first step at which the Seifert graph becomes disconnected. For this to occur, the deleted edge must have been an isthmus and the smoothed crossing must have been nugatory. This step in the construction of the resolution can be omitted since D_+ and D_- represent isotopic links and D_0 is a split link that makes zero contribution to $\nabla(z)$. Note that the homogeneity of the diagrams on the right-most branch is preserved. By pruning the resolution in this way we can ensure that the Seifert graph of the trivial link at the right-most terminal node is a tree. Thus there is at least one contribution of full rank to $\nabla(z)$. □

We have already established a relationship (Theorem 7.2.1) between the breadth of the Alexander polynomial and the genus of a link. This inequality carries over easily to the Conway polynomial. For homogeneous links, we can show that equality is achieved.

Corollary 7.6.5. *If L is a homogeneous non-split link then*

$$2\,g(L) \;=\; \mathrm{maxdeg}\,\nabla_L(z) - \mu(L) + 1.$$

PROOF. Let D be a homogeneous diagram of L, let F be the projection surface constructed from D and G be its Seifert graph. Then $\mathrm{maxdeg}\,\nabla_L(z) = \mathrm{rank}(G) = 1 - \chi(F)$. □

Corollary 7.6.6. *A projection surface constructed from a connected homogeneous diagram is a minimal-genus spanning surface for the link.*

Corollary 7.6.7. *The genus of a (p,q) torus knot with $p > q \geqslant 2$ is* $\frac{1}{2}(p-1)(q-1)$.

PROOF. Exercise 7.12.12. □

There are some knots that do not have a homogeneous diagram: any non-trivial knot K with $\nabla_K(z) = 1$ provides an example. This includes the Kinoshita–Terasaka knot, untwisted doubles (Exercise 7.12.2), and some pretzel knots (Exercise 7.12.3). It is perhaps worth highlighting one case in particular:

Corollary 7.6.8. *Non-alternating knots exist.*

PROOF. We know that the trefoil is non-trivial; hence, by Theorem 4.4.1, the untwisted double of a trefoil is non-trivial. It has Conway polynomial $\nabla(z) = 1$ so it is non-homogeneous and hence non-alternating. $\qquad\square$

7.7   Linear skein theory

We shall consider the set of all formal linear combinations of ambient isotopy classes of links with coefficients taken from a ring R. We shall take R to be the ring of polynomials in one variable, z, with integer coefficients: $R = \mathbb{Z}[z]$. An example of such an expression is

$$(3\,z)\,(4_1 \,\#\, 7_4) + (5\,z^2 - 7z + 2)\,9_{42} + (283\,z^{32})\,6_2^3 + (-78)\,0_1^4.$$

We can manipulate expressions of this type by applying the Conway skein relation as a rewriting rule. The resulting algebraic structure is called a *linear skein* and is denoted by $\mathcal{L}(\mathbb{R}^3)$.

Note that these expressions are purely formal in character — the plus operator used to form the expression does not correspond to any operation on links. An expression cannot be evaluated, it can only be replaced by an equivalent one using the skein relation.

The following example should help to underline this. It is possible to find distinct links such that switching and smoothing at a crossing site on each one leads to the same pair of links. A computer search by Morwen Thistlethwaite of knots up to 13 crossings revealed the following remarkable triplet (see Figure 7.5). The trivial link of two components is 0_1^2. Switching the indicated (positive) crossing in 13n1836 produces 10_{129} and smoothing it produces 0_1^2. Switching the indicated (negative) crossing in 10_{129} produces 8_8 and smoothing it produces 0_1^2. Now $(13n1836, 10_{129}, 0_1^2)$ and $(8_8, 10_{129}, 0_1^2)$ are both of the form (L_+, L_-, L_0) so we can write

$$13n1836 = 10_{129} + z\,0_1^2 \qquad \text{and also} \qquad 8_8 = 10_{129} + z\,0_1^2.$$

The knots 8_8 and 13n1836 are said to be *skein equivalent*. There are also examples of knots that are skein equivalent to their mirror image (see Exercise 7.12.15).

We have already seen that by repeatedly applying the skein relation we can express any element in the skein as a linear combination of trivial links. Furthermore, any terms involving split links can be replaced by zero. We can say that the skein is generated by the trivial knot.

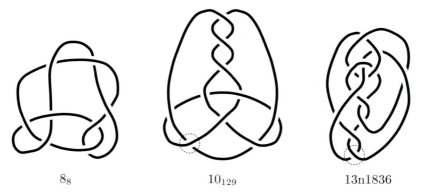

8_8 10_{129} 13n1836

Figure 7.5. The knots 8_8 and 13n1836 are skein equivalent.

Theorem 7.7.1. *Any expression in the Conway skein of oriented links in* $\mathbb{R}^3$ *is equivalent to a unique expression of the form* $p(z)\,0_1$ *where* $p(z)$ *is a polynomial in* $\mathbb{Z}[z]$.

This is just a more formal way of saying that we can resolve any link using a resolving tree with trivial links at the terminal nodes.

If we assign an element of R to the skein generator, we can evaluate any expression in the skein to produce a value in R. The choice of a value for the trivial knot is known as normalising, and the convention is to set its value to 1. The choice determines an *evaluation function* from the linear skein of ambient isotopy classes of oriented links in $\mathbb{R}^3$ to the ring of coefficients:

$$e : \mathcal{L} \longrightarrow R.$$

With this normalisation, the value assigned to an expression consisting of a single link with coefficient 1 is just its Conway polynomial: $e(1\,L) = \nabla_L(z)$.

Although this method may seem a bit formal, its strength is that it can be generalised in several ways. We can change the space in which the links are embedded, the type of objects allowed in the embedding, the ring of coefficients, and the skein relation. The last two are closely connected and cannot be chosen arbitrarily if we want the skein structure to correspond to a link invariant. The ambient space is easier to change.

Let M be an orientable 3-manifold. For example M can be $\mathbb{R}^3$, a ball, or a solid torus. If M has a boundary then we can choose $2n$ points on ∂M and label n of them as inputs and the others as outputs. An 'inhabitant' of M consists of n oriented arcs, each of which connects an input to an output, and a (possibly empty) set of loops. The loops are embedded in M, the arcs are properly embedded in M, and all these components are disjoint. The number of loops may vary (there may be none at all), but there must always be exactly n arcs.

Just as we are interested in links in $\mathbb{R}^3$ up to ambient isotopy, so we also study the ambient isotopy classes of these inhabitants: two inhabitants t_1 and t_2 are equivalent if there is an isotopy of M which is the identity on ∂M and which carries t_1 onto t_2. We denote this manifold and the set of isotopy classes of its inhabitants by $M^{\langle 2n \rangle}$. When the manifold M is a ball, $M^{\langle 2n \rangle}$ is the set of oriented n-tangles. The case $n = 0$ is equivalent to oriented links in $\mathbb{R}^3$. We shall study the cases $n = 1$ and $n = 2$.

Here we must insert a word of caution. Unfortunately, the word 'tangle' is used for many distinct but related concepts. A tangle is always a fragment of a link, but you should be clear about the following:

- the topological type of the containing 3-manifold, M;

- whether or not M has a rigid geometrical shape;

- the number, n, of inputs and outputs;

- whether the inputs and outputs have fixed locations or are free to wander over ∂M during an isotopy;

- whether loop components are allowed.

We shall now investigate the 1-tangles in a ball that may have loops. Let $\mathcal{L}(B^{\langle 2 \rangle})$ be the skein of all formal linear combinations of isotopy classes of oriented 1-tangles with coefficients taken from the ring $R = \mathbb{Z}[z]$. As before, the Conway skein relation can be applied to expressions in $\mathcal{L}(B^{\langle 2 \rangle})$ to rewrite them in an equivalent form.

Theorem 7.7.2. *Any expression in the Conway skein of oriented 1-tangles, $\mathcal{L}(B^{\langle 2 \rangle})$, is equivalent to a unique expression of the form $p(z)\,t_0$ where $p(z)$ is a polynomial in $\mathbb{Z}[z]$, and t_0 is the trivial 1-tangle.*

PROOF. By repeatedly applying the skein relation, we can express any element in the skein as a linear combination of ascending diagrams of 1-tangles. An ascending 1-tangle is a trivial 1-tangle (a ball spanned by an unknotted arc) together with a (possibly empty) set of unknotted unlinked loops. The skein relation forces any tangles with split components to go to zero. We can say that the skein is generated by the trivial tangle. □

A 1-tangle in a ball B can be closed to form a link in $\mathbb{R}^3$ by adding an arc in ∂B that connects the two endpoints of the arc. This induces a linear map from the skein $\mathcal{L}(B^{\langle 2 \rangle})$ to the skein of isotopy classes of links in $\mathbb{R}^3$:

$$\mathcal{L}(B^{\langle 2 \rangle}) \longrightarrow \mathcal{L}(\mathbb{R}^3)$$
$$p(z)\, t_0 \longmapsto p(z)\, 0_1.$$

Suppose that S is a factorising sphere for a link L in S^3, and that B_1 and B_2 are the two balls bounded by S. Then $(B_1, B_1 \cap L)$ and $(B_2, B_2 \cap L)$ are both 1-tangles. This decomposition gives a bilinear map between skeins:

$$\mathcal{L}(B^{\langle 2 \rangle}) \times \mathcal{L}(B^{\langle 2 \rangle}) \longrightarrow \mathcal{L}(\mathbb{R}^3)$$
$$(a(z)\, t_0,\, b(z)\, t_0) \longmapsto a(z)\, b(z)\, 0_1.$$

This shows that the Conway polynomial of a product of two links is the product of their polynomials: $\nabla(L_1 \# L_2) = \nabla(L_1)\, \nabla(L_2)$.

Notice that if we rotate the generating tangle t_0 by $180°$ and reverse the orientation of the arc then we cannot distinguish the result from t_0. This operation is a symmetry of the tangle. Hence if we rotate any 1-tangle by $180°$ and reverse the orientations of all its components, its expression as $p(z)\, t_0$ will be unchanged. This implies that the Conway polynomial of a link and its reverse are equal. It also means that $\nabla(L_1 \# L_2) = \nabla(L_1 \# -L_2)$, but, if neither L_1 nor L_2 is reversible, these two products are distinct links.

In the case of 1-tangles this may not seem very significant. As we shall see in the next section, the corresponding analysis of 2-tangles leads to more interesting results.

7.8 Marked tangles and mutation

In §4.9 tangles were defined as the products of a decomposition. With that definition, we have no control over the (geometric) shape of the decomposing sphere, or the location of the punctures. In some circumstances this lack of rigidity makes it possible to prove results in a very general context. However, to apply skein theory we need to work in a more restrictive category, and we consider marked tangles.

Definition 7.8.1. A *marked* 2-tangle is a round ball B that contains two disjoint properly embedded arcs, and a (possibly empty) set of disjoint embedded loops. The four ends of the arcs meet the boundary sphere in points that are equally spaced around a great circle (called the *equator*); we

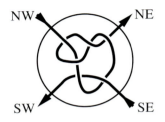

Figure 7.6. A marked tangle.

label them in sequence around the circle with NW, NE, SE and SW, and
view them so that they appear as in Figure 7.6. The arcs and loops are
oriented: each arc is directed from an input to an output, and we choose
NW and SE to be inputs and NE and SW to be outputs.

Two marked tangles are equivalent if there is an isotopy of the ball that
is the identity on the boundary of the ball, and which carries the arcs and
loops in one tangle to those in the other. In the notation of the previous
section, the set of isotopy classes of marked 2-tangles is denoted by $B^{\langle 4 \rangle}$.

We shall now construct the Conway skein of these tangles. We take the
set of all formal linear combinations of isotopy classes of tangles with coefficients in $\mathbb{Z}[z]$. As before, we use the Conway skein relation as a rewriting
rule to form the skein, $\mathcal{L}(B^{\langle 4 \rangle})$.

Theorem 7.8.2. *Any expression in the Conway skein of oriented 2-tangles,*
$\mathcal{L}(B^{\langle 4 \rangle})$, *is equivalent to a unique expression of the form* $p_0(z)\, t_0 + p_\infty(z)\, t_\infty$
where the $p_*(z)$ *are polynomials in* $\mathbb{Z}[z]$, *and* t_0 *and* t_∞ *are the tangles shown*
in Figure 7.7.

PROOF. We have to show that, by repeatedly applying the skein relation,
any element in the skein can be expressed as a linear combination of the
two tangles. The procedure is the same as for links and 1-tangles. We

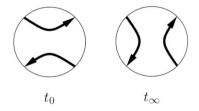

$$t_0 \qquad\qquad t_\infty$$

Figure 7.7. The generators of $\mathcal{L}(B^{\langle 4 \rangle})$.

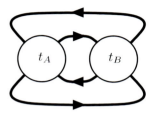

Figure 7.8. The sum of two marked tangles.

choose an ordering on the inputs to the tangle, then proceed to construct an ascending 2-tangle. Because of the arrangement of inputs and outputs around the equator, an ascending 2-tangle must be isotopic to t_0 or t_∞. As before, any tangles with split components contribute zero to the expression. The proof is completed by using induction on the number of crossings. $\square$

Theorem 7.8.3. *Suppose that a sphere meets an oriented link L in four points decomposing it into two 2-tangles. The decomposition can be arranged to look like Figure 7.8 so that L is built from two marked tangles t_A and t_B. This decomposition induces a bilinear map:*

$$\mathcal{L}(B^{\langle 4 \rangle}) \times \mathcal{L}(B^{\langle 4 \rangle}) \longrightarrow \mathcal{L}(\mathbb{R}^3)$$
$$(t_A, t_B) \longmapsto \Big(a_0(z)\, b_\infty(z) + a_\infty(z)\, b_0(z) \Big)\, 0_1$$

where $t_A = a_0(z)\, t_0 + a_\infty(z)\, t_\infty$ and $t_B = b_0(z)\, t_0 + b_\infty(z)\, t_\infty$.

PROOF. Since the map is stated in terms of generators on both sides, we need only consider how the generators are related. This is summarised in the following table. Two of the links are split and hence their contributions must be zero.

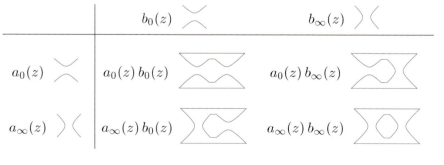

$\square$

Notice that both the generators are symmetric. A rotation of $180°$ about an axis perpendicular to the equatorial plane leaves the tangles t_0 and t_∞ unchanged. Similarly, rotations of $180°$ about the vertical North–South axis or the horizontal East–West axis, followed by reversing the orientation of the strings, are also symmetry operations. This observation provides an easy way to create links with the same Conway polynomial.

Definition 7.8.4 (mutation). Let L_1 be an oriented link which contains a marked tangle t. Remove t, rotate it by $180°$ about one of its three principal axes, reversing all orientations if necessary, and glue it back in position to form a new link L_2. If t itself has some symmetry then L_1 and L_2 may be the same. If L_1 and L_2 are different they are called *mutants* of each other.

Corollary 7.8.5. *Mutant links are skein equivalent and hence have the same Conway polynomial.*

PROOF. Let $t \in \mathcal{L}(B^{\langle 4 \rangle})$ be the tangle we use to form the mutant and let r be the rotation used to form the mutant. Tangle t can be written as $p_0(z)\, t_0 + p_\infty(z)\, t_\infty$. The rotation preserves the signs of the crossings so we can apply the same sequence of skein relations to the rotated tangle $r(t)$ to express it as $p_0(z)\, r(t_0) + p_\infty(z)\, r(t_\infty)$. But $r(t_0) = t_0$ and $r(t_\infty) = t_\infty$ so this is the same as for the original tangle t.   □

The standard example of a mutant pair is shown in Figure 7.9: on the left is the Kinoshita–Terasaka knot [146] we have met before, the other is known as the Conway knot [139]. These are well known as the smallest (lowest crossing number) non-trivial knots with Conway polynomial 1. Mutants, like skein equivalent knots in general, are difficult to distinguish. These two were first proved to be distinct using representations of the fundamental groups of their complements [152]. They also have distinct genera [142].

One way to try to distinguish mutants is to look at invariants of their satellites. Simple satellites like $(n, 2)$ cables or doubles may suffice. However, there is no point using the determinant or Conway polynomial in this way because of Theorems 6.7.7 and 7.1.4(g).

A close inspection of Figure 7.9 will show that these diagrams are not in the form of Figure 7.8: the inputs and outputs are not in the correct positions. It is simple enough to manipulate the diagrams into the correct form, but this is not necessary. The mutant construction always gives links

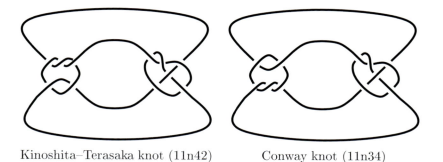

Kinoshita–Terasaka knot (11n42)   Conway knot (11n34)

Figure 7.9. Two mutant knots.

with the same polynomial however the inputs and outputs are distributed (see Exercise 7.12.14).

Mutants provide further examples of skein equivalent links. The knots 8_8 and 13n1836 are not mutants: the knot 8_8 is rational and has no mutants. So mutation and skein equivalence are distinct ideas.

7.9 Conway's fraction formula

A marked tangle can be closed to form a link in the two ways shown in Figure 7.10. The two closures will usually be distinct and may not even have the same number of components. For the tangle t shown in Figure 7.6, the denominator $D(t)$ is the knot 6_1, and the numerator $N(t)$ is the product of a Hopf link with the figure-8 knot: $2_1^2 \# 4_1$. The reason for this strange nomenclature will be revealed in the next theorem.

We can use these closures to define two evaluation functions on $\mathcal{L}(B^{\langle 4 \rangle})$:

$$\nabla_N : t \longmapsto \nabla(N(t)) \qquad \text{and} \qquad \nabla_D : t \longmapsto \nabla(D(t)).$$

These are special cases of two linear projections between skeins. For example, ∇_N comes from the map

$$\begin{array}{ccc} \mathcal{L}(B^{\langle 4 \rangle}) & \longrightarrow & \mathcal{L}(\mathbb{R}^3) \\ p_0(z)\, t_0 + p_\infty(z)\, t_\infty & \longmapsto & p_\infty(z)\, 0_1. \end{array}$$

In the tangles of §4.9, the endpoints of the arcs can move freely over the boundary sphere. This hinders attempts to define an inverse to decomposition: the ambiguity involved in any kind of composition is much greater than that for the product operation. However, with marked tangles, various well-defined composition operations are possible.

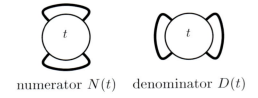

numerator $N(t)$ denominator $D(t)$

Figure 7.10. The two ways to close a marked tangle.

The *partial sum* of two marked tangles t_A and t_B is produced as shown in Figure 7.11 and is denoted by $t_A \oplus t_B$. This operation is associative but not commutative. Note that it may generate an extra loop component.

The denominator of a partial sum is just a product of two oriented links:

$$D(t_A \oplus t_B) \;=\; D(t_A) \,\#\, D(t_B).$$

The numerator of a partial sum $N(t_A \oplus t_B)$ is called the *sum* of the tangles and corresponds to the decomposition of a link into tangles shown in Figure 7.8.

Theorem 7.9.1 (Conway). *Given marked 2-tangles t_A and t_B,*

(a) $\nabla_N(t_A \oplus t_B) \;=\; \nabla_N(t_A)\,\nabla_D(t_B) + \nabla_D(t_A)\,\nabla_N(t_B),$

(b) $\nabla_D(t_A \oplus t_B) \;=\; \nabla_D(t_A)\,\nabla_D(t_B).$

PROOF. Part (a) is just a restatement of Theorem 7.8.3. Part (b) is the formula for the product of links. □

Now we can see the reason for the numerator–denominator labels.

Corollary 7.9.2. *Let $F(t) = \nabla_N(t)/\nabla_D(t)$. We call this quotient the fraction of a tangle t. It is a purely formal expression — we are not allowed to cancel factors. If $t_A \neq t_\infty \neq t_B$ then*

$$F(t_A \oplus t_B) \;=\; F(t_A) + F(t_B).$$

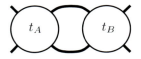

Figure 7.11. The partial sum of two marked tangles: $t_A \oplus t_B$.

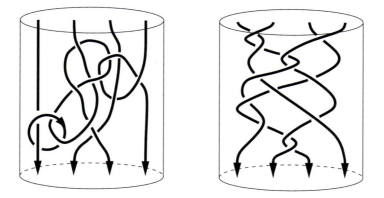

Figure 7.12. Two 4-tangles. The one on the right is also a 4-string braid.

7.10 Cylindrical tangles

For tangles with many components it is helpful to specify a geometry of the points on the boundary of M as we did for marked 2-tangles. Instead of having the inputs and outputs alternate, as they do around the equator in Figure 7.6, we shall collect all the endpoints of each type together.

A cylindrical n-tangle is a ball $D^2 \times [0,1]$ with a line of n inputs in the top $(D^2 \times 1)$ and a line of n outputs directly below them at the bottom. As usual, an inhabitant is a set of n arcs oriented from top to bottom plus a (possibly empty) set of loops. Two examples are shown in Figure 7.12. As usual, we consider tangles up to equivalence using ambient isotopy classes of inhabitants with fixed endpoints.

The linear skein $\mathcal{L}(B^{\langle 2n \rangle})$ of these n-tangles is generated by $n!$ elements, which can be thought of as permutations of n points (Exercise 7.12.17). These tangles have a composition rule: stack two cylinders on top of each other to produce another. This induces a map from $\mathcal{L}(B^{\langle 2n \rangle})$ to itself. We can also *close* a tangle by gluing the top of the cylinder to the bottom, thus forming a link. As before, this induces a map from $\mathcal{L}(B^{\langle 2n \rangle})$ to $\mathcal{L}(\mathbb{R}^3)$.

One particular class of these tangles has received much attention. Notice that the tangle on the right of Figure 7.12 is simpler than the one on the left: its strings are just 'braided' rather than tangled up. Each disc $D^2 \times x$ in the cylinder meets the inhabitant in exactly n points, once in each string; there are no extra loops. Such a tangle is called an n-string braid. Its strings descend monotonically without reversing anywhere. Braids are equivalent if they are isotopic using braid-preserving isotopies so that all the intervening

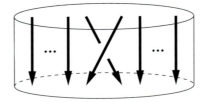

Figure 7.13. The elementary braid σ_i: the $(i+1)$th string crosses over the ith string.

positions are also braids. In fact, this is no restriction since braids that are equivalent as tangles are also braid equivalent [138].

Any link can be presented as the closure of an n-tangle for any $n \geqslant 0$. Braids are more constraining and it is not immediately clear that every link can be presented as a closed braid. We shall see that this is indeed the case in Chapter 10.

Braids were introduced in the 1920s by Emil Artin. The n-string braids form a group (Exercise 7.12.18), denoted by $\mathbb{B}_n$, and this allows some algebraic methods to be used in knot theory. The group is generated by $n-1$ *elementary* braids, each of which contains a single crossing. The generator[*] σ_i is shown in Figure 7.13. Its inverse, σ_i^{-1}, has a crossing of opposite sign. Writing down words in these braid generators provides a way to describe links without the use of pictures. For example $\sigma_1 \sigma_2^{-1} \sigma_1 \sigma_2^{-1} \in \mathbb{B}_3$ closes to give the figure-8 knot. This allows easy input of a link description into a computer.

The fact that braids have an algebraic structure provides another route to link invariants. What is required is a function from the braid group to another mathematical structure which gives the same value on braids that close to the same link. A significant discovery of this kind was made in the early 1980s by Vaughan Jones [182]: we shall learn more about it in Chapter 9.

The braid approach to link invariants also provides a means to speed up the computation of linear skein invariants like the Conway polynomial. For a diagram with n crossings, the time taken to compute a polynomial using the resolving tree type of algorithm grows exponentially like 2^n. For a fixed

---

[*]The algebraic theory of braids developed fairly independently of topologically based knot theory. One unfortunate consequence of this is that the original 'positive' generator had a negative crossing and its inverse had a positive crossing. This convention is beginning to be reversed; the more consistent approach is used here.

number of braid strings, an algorithm based on representation theory can be developed that grows only polynomially in the number of crossings — at worst n^3 [149, 150, 255]. It grows factorially as the number of braid strings increases, but many large knots can be presented as braids with few strings. With today's machines, this is less of an issue, but in the late 1980s, when a 'powerful' computer had 16 MB of memory and a 33 MHz clock, it made the difference between being able to compute polynomials of simple cable knots and not.

7.11 A troublesome pair

In Exercise 6.9.5 you found that knots 6_1, 8_{20} and the reef knot $3_1 \# 3_1^*$ all have determinant 9 and signature 0. The techniques of this chapter discriminate 6_1 from the others, but 8_{20} and the reef knot have the same Conway polynomial. This means that we still cannot complete the classification of the knots shown at the start of the book in Figure 1.6.

The table in Figure 7.14 lists some properties of these two knots, roughly in the order that they are introduced in the book. The values marked † are ones that cannot be established with the methods seen so far. Unfortunately, these are the only ones that separate the two knots.

Here are the reasons for some of the values given.

- The computable invariants (det, σ and $\nabla(z)$) are all equal, so any

Property	$3_1 \# 3_1^*$	8_{20}
Reversible	yes	yes
Amphicheiral	yes	no†
Unknotting number	$2^†$	1
Polygon index	8	8
Crossing number	6	$8^†$
Alternating	yes	no†
Prime	no	yes†
Ribbon	yes	yes
Bridge number	$3^†$	$3^†$
Genus	2	2
Determinant	9	9
Signature	0	0
Conway polynomial	$(1 + z^2)^2$	$1 + 2z^2 + z^4$

Figure 7.14. Invariants and other properties of two knots from Figure 1.6.

consequences derived from them will be no use in distinguishing the two knots. For instance, we can prove a knot is cheiral if it has non-zero signature but 8_{20} has the same signature as an amphicheiral knot.

- For the two granny knots, the bound obtained from the signature is sufficient to prove that the unknotting number must be 2, but this trick does not work for the reef knot. All we can say is that $1 \leqslant u(3_1 \,\#\, 3_1^*) \leqslant 2$. Experiment shows that $u(8_{20}) = 1$.

- We can distinguish $3_1 \,\#\, 3_1^*$ and 8_{20} from all the knots that can be made with fewer than eight edges, hence they both have polygon index 8. The crossing number of $3_1 \,\#\, 3_1^*$ can be established in the same way, but not $c(8_{20})$.

- Our tests for primality are rather limited: genus-1 knots, torus knots, rational knots, and certain satellites. None of these applies to 8_{20}.

- The reef knot is a ribbon knot by Exercise 5.9.9, and Figure 4.8 shows a ribbon presentation of 8_{20}.

- The genus is bounded below by half the degree of the Conway polynomial and it is easy to find genus-2 spanning surfaces for both knots.

Our study of knottedness has progressed a long way, and we can prove many things about large families of knots. However, as this example shows, it can be extremely difficult to establish specific properties of a particular knot or link you are interested in.

7.12 Exercises

1. The figure below shows an untwisted double of the figure-8 knot. Using the obvious surface, write down a Seifert matrix and compute the Alexander polynomial.

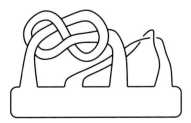

2. Show that $\Delta_K(t) = 1$ whenever K is an untwisted double.

3. Using the Seifert matrix you worked out in Exercise 6.9.6, find a formula for the Alexander polynomial of the (p, q, r) pretzel knot with p, q and r all odd.

4. Which pretzel knots have $\Delta(t) = 1$? Compare your answer with that to Exercise 6.9.7.

5. Calculate the Conway polynomial of the (p, q, r) pretzel knot with p, q and r all odd using the skein relation.

6. Show that if K is a knot then the constant term of $\nabla(K)$ is 1. (Hint: what is the consequence of putting $z = 0$ in the skein relation?)

7. Show that if L is a link with μ components then $\nabla(L^*) = (-1)^{\mu-1}\nabla(L)$. Deduce that the Whitehead link is cheiral.

8. Complete the proof of Proposition 7.4.3.

9. Let K be the knot shown in Figure 7.3. Show that

(a) K is the trivial knot if and only if $a_n = 0$ and $a_i = 1$ for all $i < n$,

(b) appending $(\overbrace{1, \dots, 1}^{m}, 0)$ to any sequence where $m \geqslant 0$ does not change the polynomial or the knot,

(c) if $a_n \neq 0$ then K has genus n,

(d) if K is non-trivial then it has unknotting number 1.

10. An almost positive link is one that has a diagram in which every crossing except one is positive. Show that an almost positive link has a positive Conway polynomial. (Choose a basepoint and ordering carefully.)

11. Show that an alternating diagram is also a homogeneous diagram.

12. Prove Corollary 7.6.7.

13. The diagram of 10_{50} shown on the right of Figure 1.3 can be decomposed by a horizontal line into two tangles with five crossings in each. Redraw this decomposition in the form of Figure 7.8. Show that this splitting does not produce mutants — all the rotations give 10_{50}.

14. Fill in the details of the proof of Theorem 7.8.2.

15. The knot shown below is a 12-crossing diagram of 10_{48}. This knot can be shown to be cheiral via properties of its Kauffman polynomial (see page 238), and it is also reversible. By using the circled crossings, show that the knot is skein equivalent to (the reverse of) its mirror image [187].

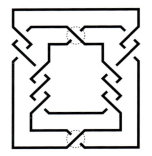

16. Suppose that we change the ordering of the inputs and outputs in the definition of a marked tangle: now NW and SW are inputs, and NE and SE are outputs. Show that the Conway skein of these tangles is generated by the two tangles below.

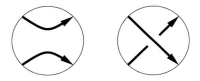

Hence show that the mutants in Figure 7.9 have the same Conway polynomial.

17. Show that the linear skein $\mathcal{L}(B^{\langle n \rangle})$ of n-tangles has $n!$ generators. Each can be thought of as giving a permutation of the points $1, \ldots, n$ with the arcs placed in layers so that the arc with input i is in layer i.

18. Consider the set $\mathbb{B}_n$ of n-string braids under the composition rule given by stacking cylinders: to form $\beta_1 \beta_2$ place braid β_1 above β_2.

(a) Show that $\mathbb{B}_n$ is a group.

(b) Show that the elementary braids σ_i generate $\mathbb{B}_n$.

(c) Show that $\sigma_i \sigma_j = \sigma_j \sigma_i$ when $|i - j| \geqslant 2$.

(d) Show that $\sigma_i \sigma_{i+1} \sigma_i = \sigma_{i+1} \sigma_i \sigma_{i+1}$.

8 Rational Tangles

At the end of Chapter 4, we saw how links could be decomposed into fragments called tangles. In particular, rational links have a decomposition into two trivial 2-tangles. In the last chapter we introduced a more rigid framework for studying 2-tangles using marked tangles. We now investigate the marked tangles that can be used to build rational links. This leads to a particularly nice classification scheme and a close relationship with the rational numbers, whence the name.

8.1 Generating rational tangles

A rational tangle is homeomorphic to the trivial 2-tangle. This means that it can be 'unwound' by sliding the endpoints around on the boundary sphere. We want to use this idea in reverse, starting with a basic object and applying a sequence of operations to it to build up complex-looking tangles.

Let us first define four operations that can be performed on a marked tangle (see Figure 8.1):

H Take the two right-hand ends (NE and SE) and twist them so that the over-crossing strand created has positive gradient.

V Take the two lower ends (SW and SE) and twist them so that the over-crossing strand created has positive gradient.

R The points NW and SE determine a line through the tangle. Using this line as an axis, rotate the tangle 180°.

F The points NW and SE determine a plane through the tangle that is orthogonal to the equatorial plane. Reflect the tangle in this plane. (Recall that the equatorial plane is the one containing the four endpoints.)

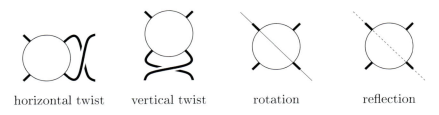

horizontal twist   vertical twist   rotation   reflection

Figure 8.1. Operations on rational tangles.

The first three operations are direct and can be used to construct a tangle from two pieces of string. Operation F is indirect as it involves taking a reflection. Although this 'virtual operation' cannot be easily carried out on a physical example of a tangle, its inclusion in the list simplifies the mathematical analysis of the situation.

It may appear from the description that H and V add twists of the same sense, but this is not in fact the case. Repeating operation H will build up a twist-box containing positive half-twists (in the sense of Figure 3.7), while repeating operation V builds up negative half-twists.

These operations are applied to tangles and we shall write the result of applying operation H to tangle t as $H(t)$ in the usual way. We use the tangles t_0 and t_∞ shown in Figure 8.2 as our starting materials. Clearly $H(t_\infty) = t_\infty$, $V(t_0) = t_0$, $R(t_\infty) = t_0 = F(t_\infty)$ and $R(t_0) = t_\infty = F(t_0)$. We shall denote the compound operation of H repeated n times by H^n, and use H^{-1} to denote a horizontal twist in the opposite sense. Similarly for the other operations. Once we start composing operations, we need the identity operation: this will be denoted by I. A little experimentation shows that

- $R^2 = F^2 = I$,
- $V^{-1} = RHR$,
- $V = HRH$,
- $V = FHF$.

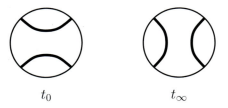

t_0 t_∞

Figure 8.2. The two simplest rational tangles.

John Conway's algorithm for constructing rational tangles is simple to describe. Its input is a sequence of integers a_1, a_2, ..., a_n. With the exception of a_1, all the coefficients can be assumed to be non-zero. (The symbol ':=' is an assignment operator — the object on the left takes the value on the right.)

$t := t_\infty$

for each i from n down to 1 do
{
$\quad\quad t := F(t)$
$\quad\quad t := H^{a_i}(t)$
}

In words, this says take tangle t_∞, perform operation F, then add a_n horizontal twists, perform operation F again, then add a_{n-1} horizontal twists, continuing until all the coefficients have been used. Note that i counts down from n to 1. The resulting tangle is denoted by $C(a_1, a_2, \ldots, a_n)$. The stages in the construction of the tangle corresponding to the sequence $C(3, 2, 5, 4)$ are shown in Figure 8.3. As F is a reflection, the sign of the twists changes at each step.

Although this process is simple to describe, it is of no practical use because we cannot change the sign of a twist-box at will. Looking at the finished tangle, it is easy to see that $C(3, 2, 5, 4)$ can be constructed as $H^3 V^2 H^5 V^4 (t_\infty)$. As H and V produce twist-boxes of opposite signs, the exponents are all positive.

Recasting the HF algorithm in terms of the horizontal and vertical twist operations (H and V) forces us to separate the process into two cases according to the parity of n: the construction always ends with a horizontal twist-box, so we must start with a V if n is even, and an H if n is odd. Thus we get

$$H^{a_1} V^{a_2} \ldots V^{a_{n-1}} H^{a_n} (t_0) \quad \text{if } n \text{ is odd, and}$$
$$H^{a_1} V^{a_2} \ldots H^{a_{n-1}} V^{a_n} (t_\infty) \quad \text{if } n \text{ is even.}$$

The tangles produced in these two cases are shown in Figure 8.4 — the twist-boxes have been aligned to save space.

This generating mechanism can produce the same result starting from different sequences. For example $H V^{-1} H^2 V^{-1} H(t_0) = t_0$. One of our objectives is to explain which sequences produce isotopic tangles.

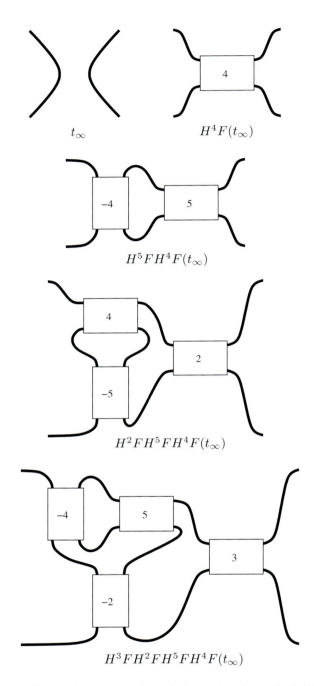

t_∞ $H^4F(t_\infty)$

$H^5FH^4F(t_\infty)$

$H^2FH^5FH^4F(t_\infty)$

$H^3FH^2FH^5FH^4F(t_\infty)$

Figure 8.3. Conway's construction of the rational tangle $C(3, 2, 5, 4)$.

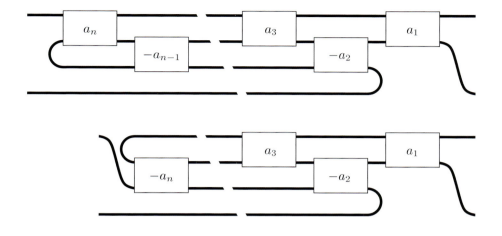

Figure 8.4. Diagrams of rational tangles denoted by $C(a_1, a_2, \ldots, a_n)$. The format depends on whether n is odd or even.

8.2 Classification of rational tangles

We shall classify rational tangles by relating them to loops on a torus. On the left of Figure 8.5 is a torus, T, together with an axis. The axis punctures the torus in four places and the configuration is arranged so that a rotation of $180°$ about the axis carries the torus to itself. The rotation induces a homeomorphism $\rho : T \longrightarrow T$.

Define the relation $\sim$ on T by $x \sim y$ if and only if $x = y$ or $\rho(x) = y$. It is an equivalence relation because ρ^2 is the identity. We can define a new space, denoted by $T/\sim$ or T/ρ, in which related points are identified: each point in $T/\sim$ is an equivalence class $\{x, \rho(x)\}$ and the whole space is created by taking all points $x \in T$. The map $\phi : T \longrightarrow T/\sim$ defined as $x \longmapsto \{x, \rho(x)\}$ is called the *projection* map. As T is a topological space, the quotient space $T/\sim$ can be given the quotient topology: a set $U \subset T/\sim$ is open if and only if the preimage $\phi^{-1}(U)$ is open in T.

The sequence of pictures in Figure 8.5 shows that $T/\sim$ is, in fact, a sphere: it is a cylinder (half the torus) with each end sewn up, forming something like a pillowcase. It is important to note that the four corners of the pillowcase are different from the other points: they are the projections of the places where the axis punctures the torus and are called *branch points*. Each of the other points in $T/\sim$ has two preimages in T, but the branch points come from the fixed points of ρ and so each one has a single preimage.

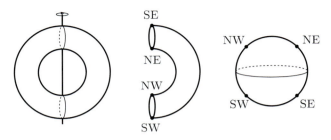

Figure 8.5. A torus is a 2-fold branched cover of a sphere.

If you are familiar with covering spaces, you will recognise this configuration: T is a 2-fold branched cover of $T/\sim$ with four branch points, T is the covering space, $T/\sim$ is the base space, and ρ is the covering involution. We shall call $T/\sim$ a 4-point sphere.

Suppose that $(B, \alpha_1 \cup \alpha_2)$ is a rational tangle. Identify ∂B with the 4-point sphere so that the endpoints of the arcs α_1 and α_2 lie on the four branch points. By Lemma 4.10.2, we can isotop the arcs in B, keeping their endpoints fixed, until they lie in ∂B. The preimages $\phi^{-1}(\alpha_1 \cup \alpha_2)$ of the arcs are two closed curves in the torus. In fact, they are two parallel copies of some torus knot $T(p, q)$. An example is shown in Figure 8.6: part (a) is the tangle $C(2, 3)$ with the two arcs distinguished; in part (b) the ball has been deformed into a solid cylinder and the two arcs isotoped to lie in its boundary; in part (c) the arcs in ∂B have been lifted to two copies of the $(7, 3)$ torus knot in the covering torus.

So to every rational tangle we can associate a loop on the torus, and hence a pair of coprime integers. A rational tangle which lifts to the (p, q) torus knot is called the p/q tangle and denoted by $t_{p/q}$. The two simplest rational tangles, shown in Figure 8.2, lift to the longitude $T(0, 1)$ and meridian $T(1, 0)$ of T. These tangles are given the symbols t_0 and t_∞ respectively. We shall show that Conway's algorithm can construct a tangle for each value in the extended rationals $\mathbb{Q} \cup \{\infty\}$.

When the arcs of a rational tangle are made to lie on its boundary, the tangle is sometimes said to be in *pillowcase* form. The tangle $t_{7/3}$ from Figure 8.6(b) is redrawn in Figure 8.7. The values of p and q can be read from the figure by counting the gaps between the threads (seven across the top and three down each side); the sign of the fraction is positive since the strings lying on the upper side of the pillowcase have positive gradient. If we rotate the tangle labelled p/q by $90°$, we obtain the tangle with fraction

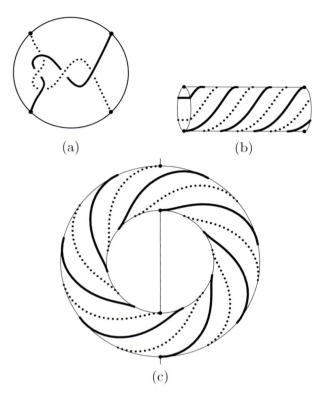

(a)

(b)

(c)

Figure 8.6. A rational tangle can be isotoped to lie on the boundary, then lifted to the covering torus.

$-q/p$. To get the *reciprocal* tangle $t_{q/p}$ we have to rotate by $90°$ and reflect in the plane of the picture to switch the crossings.

Using these observations we can see that the tangle $t_{7/3}$ in Figure 8.6(a) is the partial sum $t_{1/3} \oplus t_{2/1}$. The fact that $7/3 = 2 + 1/3$ is more than a co-

Figure 8.7. The rational tangle $t_{7/3}$ in 'pillowcase' form.

incidence, and we can define a tangle arithmetic that mirrors the arithmetic in $\mathbb{Q}$. This is why the rational tangles are labelled with fractions rather than called (p, q) tangles. The next four sections of this chapter are devoted to a discussion that leads up to an explanation of this magic.

Before moving on we make the following observation.

Theorem 8.2.1. *Rational tangles cannot be used to form mutants.*

PROOF. The pillowcase form of a rational tangle exhibits a lot of symmetry. Rotating it by $180°$ about any of the principal axes (North–South, East–West, or orthogonal to the equatorial plane) produces the same tangle. Therefore, performing a mutation operation with a rational tangle will not change the link.   □

8.3   Homology and maps

The operations used to build rational tangles can be seen as homeomorphisms of the ball to itself. Our goal is to represent these maps by matrices, which are more amenable to study. For this purpose we need to make another short excursion into homology theory.

Let X be a simplicial complex and let $h : X \longrightarrow X$ be a homeomorphism that maps simplices to simplices. We can define a homomorphism on the group of n-chains of X as follows:

$$
\begin{aligned}
h_{*n} : C_n(X) \;\;&\longrightarrow\;\; C_n(X) \\
\sum \lambda_i \, \sigma_i \;\;&\longmapsto\;\; \sum \lambda_i \, h(\sigma_i).
\end{aligned}
$$

This map is defined for all n. When the dimension can be understood from the context, we sometimes just write h_*. The two following theorems show that h_* preserves both cycles and boundaries.

Theorem 8.3.1. *The map h_* sends 1-cycles to 1-cycles.*

PROOF. Recall that a 1-simplex (or edge) in X is the image under a linear map of the standard 1-simplex — the line segment connecting $(1, 0)$ to $(0, 1)$. Therefore the map is completely determined by its value on the endpoints. Hence, writing the edge as $[v_i, v_j]$, we get

$$
\partial_1 \circ h \, [v_i, v_j] \;=\; \partial_1 [h(v_i), h(v_j)] \;=\; h(v_j) - h(v_i) \;=\; h \circ \partial_1 \, [v_i, v_j],
$$

that is, the boundary of the image is the image of the boundary.

A 1-cycle is an element z of $C_1(X)$ such that $\partial_1(z) = 0$.

$$
\begin{aligned}
(\partial_1 \circ h_{*1})(z) &= (\partial_1 \circ h_{*1})\left(\sum \lambda_i\, e_i\right) \\
&= \partial_1\left(\sum \lambda_i\, h(e_i)\right) \\
&= \sum \lambda_i\, \partial_1(h(e_i)) \\
&= \sum \lambda_i\, h(\partial_1(e_i)) \\
&= h_{*0}\left(\sum \lambda_i\, \partial_1(e_i)\right) \\
&= (h_{*0} \circ \partial_1)\left(\sum \lambda_i\, e_i\right) \\
&= (h_{*0} \circ \partial_1)(z).
\end{aligned}
$$

Since $\partial_1(z) = 0$ and h_{*0} is a homomorphism, we have $\partial_1 \circ h_*(z) = 0$. $\square$

Theorem 8.3.2. *The map h_* sends 1-boundaries to 1-boundaries.*

PROOF. This proceeds in the same way as the last theorem. A 2-simplex is determined by its three vertices so $\partial_2 \circ h = h \circ \partial_2$ and consequently $\partial_2 \circ h_{*2} = h_{*1} \circ \partial_2$ as before.

A 1-boundary is an element z of $C_1(X)$ such that there exists a 2-chain $t \in C_2(X)$ with $\partial_2(t) = z$. Now $h_*(z) = h_* \circ \partial_2(t) = \partial_2 \circ h_*(t)$, hence $h_*(z)$ is also a boundary. $\square$

These two results mean that, in fact, h_* passes down to the quotient and is also a map between the homology groups:

$$
h_* : H_1(X) \longrightarrow H_1(X).
$$

This construction has the following useful property.

Theorem 8.3.3. *Let g and h be homeomorphisms of X. Then $(g \circ h)_* = g_* \circ h_*$.*

PROOF. This is a straightforward calculation:

$$
\begin{aligned}
(g \circ h)_*\left(\sum \lambda_i\, e_i\right) &= \sum \lambda_i\, g(h(e_i)) \\
&= g_*\left(\sum \lambda_i\, h(e_i)\right) \\
&= g_* \circ h_*\left(\sum \lambda_i\, e_i\right).
\end{aligned}
$$

$\square$

Corollary 8.3.4. *The map $h_* : H_1(X) \longrightarrow H_1(X)$ is a group isomorphism.*

PROOF. The identity homeomorphism on X maps to the identity transformation on $H_1(X)$. Since h is a homeomorphism of X, h^{-1} exists and is also a homeomorphism. By the preceding theorem, $(h_*)^{-1} = (h^{-1})_*$ and so h_* is invertible and therefore a bijection. $\qquad\qquad\square$

8.4   Automorphisms of the torus

A homeomorphism from a space X to itself is called an *automorphism* of X. The set of all automorphisms of X is a group under composition of maps, and is denoted by $\mathrm{Aut}(X)$.

To classify the rational tangles we first need to understand the automorphisms of the torus. In §1.5 we parametrised the torus using angles (θ, ϕ) as a coordinate system, for θ and ϕ in $[0, 2\pi]$. We noted that the curves $\theta = 0$ and $\phi = 0$ are the longitude and meridian, respectively.

In the following maps, we perform addition modulo 2π:

$$
\begin{aligned}
\text{inversion} &\quad : (\theta, \phi) \longmapsto (\phi, \theta), \\
\text{meridional twist} &\quad : (\theta, \phi) \longmapsto (\theta + \phi, \phi), \\
\text{longitudinal twist} &\quad : (\theta, \phi) \longmapsto (\theta, \phi + \theta).
\end{aligned}
$$

The first map interchanges the longitude and meridian, and reverses the orientation of the torus in doing so. The other two maps, called *Dehn twists*, are orientation-preserving. It is clear that all three maps are continuous and invertible. Hence they are homeomorphisms of the torus.

Now let us apply the results of the previous section.[*] The homology group of the torus T is generated by two cycles and we can take these to be the meridian m and the longitude l. The homology group is then

$$
H_1(T) \;=\; \mathbb{Z} \oplus \mathbb{Z} \;=\; \langle m \rangle \oplus \langle l \rangle.
$$

The image of a cycle under a homeomorphism h is another cycle so we can write $h(m) = am + bl$ and $h(l) = cm + dl$ for some integers a, b, c and d.

---

[*]Strictly speaking, we need the torus to be triangulated so that it is a simplicial complex, and we need the automorphisms to preserve the simplicial structure. We can subdivide the simplices so that any given homeomorphism can be approximated arbitrarily closely by a piecewise linear one.

By choosing basis vectors for m and l, we can express h in matrix form:

$$m = \begin{pmatrix} 1 \\ 0 \end{pmatrix}, \quad l = \begin{pmatrix} 0 \\ 1 \end{pmatrix}, \quad h = \begin{pmatrix} a & c \\ b & d \end{pmatrix}.$$

The matrix has integer entries, and is invertible since h is invertible. This means the matrix belongs to the general linear group $\mathrm{GL}(2, \mathbb{Z})$ and has determinant ± 1.

Let $\Phi : \mathrm{Aut}(T) \longrightarrow \mathrm{GL}(2, \mathbb{Z})$ be the map that takes $h : T \longrightarrow T$ to the matrix representation of $h_* : H_1(T) \longrightarrow H_1(T)$. The homeomorphisms defined above are sent to the following matrices:

$$\begin{pmatrix} 0 & 1 \\ 1 & 0 \end{pmatrix}, \quad \begin{pmatrix} 1 & 1 \\ 0 & 1 \end{pmatrix}, \quad \begin{pmatrix} 1 & 0 \\ 1 & 1 \end{pmatrix}.$$

The group $\mathrm{GL}(2, \mathbb{Z})$ is generated by these three matrices (Exercise 8.10.10). This means that Φ is surjective.

We note in passing that any homeomorphism can be deformed by an isotopy. Two homeomorphisms that are isotopic will be sent to the same matrix, hence Φ is not injective. The kernel of Φ consists of all the homeomorphisms that are isotopic to the identity map. The quotient group $\mathrm{Aut}(T)/\ker(\Phi)$ is called the *mapping class group* of the torus and is isomorphic to $\mathrm{GL}(2, \mathbb{Z})$.

8.5 Representing tangle operations by matrices

The four operations defined in §8.1 can be thought of as automorphisms of the 4-point sphere bounding the tangle. Such automorphisms lift to automorphisms of the covering torus in a natural way. Furthermore, composition of tangle-generating operations lifts to composition of automorphisms on the torus, which, in turn, corresponds to multiplication of matrices in $\mathrm{GL}(2, \mathbb{Z})$.

Operations H and V lift to the meridional and longitudinal twists on the torus, and F lifts to the inversion map. The lift of operation R interchanges the longitude and the meridian sending the longitude to the meridian, and the meridian to the reverse of the longitude: this map preserves orientation. Therefore, the four operations correspond to the following matrices:

$$H = \begin{pmatrix} 1 & 1 \\ 0 & 1 \end{pmatrix}, \quad V = \begin{pmatrix} 1 & 0 \\ 1 & 1 \end{pmatrix}, \quad R = \begin{pmatrix} 0 & 1 \\ -1 & 0 \end{pmatrix}, \quad F = \begin{pmatrix} 0 & 1 \\ 1 & 0 \end{pmatrix}.$$

We saw that the tangle $C(3, 2, 5, 4)$ is generated using the following sequence of operations: $H^3 V^2 H^5 V^4$. We can represent this construction and its result as a matrix product:

$$\begin{pmatrix} 1 & 3 \\ 0 & 1 \end{pmatrix} \begin{pmatrix} 1 & 0 \\ 2 & 1 \end{pmatrix} \begin{pmatrix} 1 & 5 \\ 0 & 1 \end{pmatrix} \begin{pmatrix} 1 & 0 \\ 4 & 1 \end{pmatrix} = \begin{pmatrix} 159 & 38 \\ 46 & 11 \end{pmatrix}.$$

The intermediate matrices are

$$\begin{pmatrix} 1 & 0 \\ 0 & 1 \end{pmatrix} \xrightarrow{V^4} \begin{pmatrix} 1 & 0 \\ 4 & 1 \end{pmatrix} \xrightarrow{H^5} \begin{pmatrix} 21 & 5 \\ 4 & 1 \end{pmatrix} \xrightarrow{V^2} \begin{pmatrix} 21 & 5 \\ 46 & 11 \end{pmatrix} \xrightarrow{H^3} \begin{pmatrix} 159 & 38 \\ 46 & 11 \end{pmatrix}.$$

Clearly, distinct automorphisms of the 4-point sphere will lift to distinct automorphisms of the torus. The converse is not true. For example, the covering involution ρ is a non-trivial automorphism of the torus that reverses the orientations of both the longitude and the meridian. But under the covering projection, it becomes the identity map on the 4-point sphere. In fact, any torus automorphism h will be indistinguishable from $h \circ \rho$ after projection.

The matrix corresponding to ρ is $-I$ where I is the identity in $\mathrm{GL}(2, \mathbb{Z})$. To obtain the automorphism group of the 4-point sphere we need to add the relation $I \sim -I$ to $\mathrm{GL}(2, \mathbb{Z})$. The quotient $\mathrm{GL}(2, \mathbb{Z})/\{\pm I\}$ is called the projective general linear group and is denoted by $\mathrm{PGL}(2, \mathbb{Z})$.

An automorphism can either preserve or reverse orientation. This can be detected in the associated matrix using the sign of the determinant: $+1$ for orientation-preserving and -1 for orientation-reversing. The subgroup of $\mathrm{GL}(2, \mathbb{Z})$ formed by taking the matrices of determinant $+1$ is called the special linear group $\mathrm{SL}(2, \mathbb{Z})$. The group $\mathrm{PSL}(2, \mathbb{Z})$, often called the *modular group*, is defined in the same way.

The connections with matrix groups can be summarised as follows:

$\mathrm{GL}(2, \mathbb{Z})$,   automorphisms of the torus;
$\mathrm{SL}(2, \mathbb{Z})$,   orientation-preserving automorphisms of the torus;
$\mathrm{PGL}(2, \mathbb{Z})$, automorphisms of the 4-point sphere;
$\mathrm{PSL}(2, \mathbb{Z})$, orientation-preserving automorphisms of the 4-point sphere.

Let us now investigate the properties of the matrices associated with the tangle-generating operations. The matrices H, V and R belong to $\mathrm{PSL}(2, \mathbb{Z})$ but $\det(F) = -1$, corresponding to the fact that F reverses orientation. In the projective groups, we cannot distinguish between matrices that differ only in sign. Therefore $R^2 = -I = I$ and $R = -R = R^{-1}$. Besides I and

R, the modular group has only two other elements of finite order. Letting $P = HR$ and using the relations $V^{-1} = RHR$ and $V = HRH$ we see that

$$P^3 = (HR)^3 = (HRH)(RHR) = VV^{-1} = I.$$

Hence P and $P^2 = P^{-1}$ both have order 3.

We know that any rational tangle can be constructed using operations H and V. This is just another way of saying that $\mathrm{PSL}(2, \mathbb{Z})$ is generated by matrices H and V. However, the group is not freely generated: different products of H and V can give the same matrix. If we can understand which ones then we can determine which tangles are isotopic.

Any rational tangle can be expressed as a sequence containing H and any one of V, R and F. We can interconvert between these quite easily. Since $F^2 = I$ and $V = FHF$, we have $V^n = FH^n F$. We can do a similar thing remaining strictly within $\mathrm{PSL}(2, \mathbb{Z})$ by using R in place of F but then we need to worry about signs. To simplify some expressions we shall introduce another matrix: $M(n) = H^n F$. We can use this to rewrite a product of H's and V's in terms of M:

$$H^{a_i} V^{a_{i+1}} = H^{a_i} F H^{a_{i+1}} F = M(a_i) M(a_{i+1}).$$

We have the following equivalences among expressions.

Case 1: $H^{a_1} V^{a_2} \ldots V^{a_{n-1}} H^{a_n}$

$\equiv H^{a_1} R H^{-a_2} R \ldots R H^{-a_{n-1}} R H^{a_n}$

$\equiv H^{a_1} F H^{a_2} F \ldots F H^{a_{n-1}} F H^{a_n}$

$\equiv M(a_1) M(a_2) \ldots M(a_{n-1}) M(a_n) F.$

Case 2: $H^{a_1} V^{a_2} \ldots H^{a_{n-1}} V^{a_n}$

$\equiv H^{a_1} R H^{-a_2} R \ldots H^{a_{n-1}} R H^{-a_n} R$

$\equiv H^{a_1} F H^{a_2} F \ldots H^{a_{n-1}} F H^{a_n} F$

$\equiv M(a_1) M(a_2) \ldots M(a_{n-1}) M(a_n).$

Case 3: $V^{a_1} H^{a_2} \ldots V^{a_{n-1}} H^{a_n}$

$\equiv R H^{-a_1} R H^{a_2} \ldots R H^{-a_{n-1}} R H^{a_n}$

$\equiv F H^{a_1} F H^{a_2} \ldots F H^{a_{n-1}} F H^{a_n}$

$\equiv M(0) M(a_1) M(a_2) \ldots M(a_{n-1}) M(a_n) F.$

Case 4: $V^{a_1} H^{a_2} \ldots H^{a_{n-1}} V^{a_n}$

$\equiv \; R H^{-a_1} R H^{a_2} \ldots H^{a_{n-1}} R H^{-a_n} R$

$\equiv \; F H^{a_1} F H^{a_2} \ldots H^{a_{n-1}} F H^{a_n} F$

$\equiv \; M(0) M(a_1) M(a_2) \ldots M(a_{n-1}) M(a_n).$

We can regard case 3 as a special form of case 1 obtained by setting $a_1 = 0$. Similarly, case 4 is a special form of case 2.

8.6 Continued fractions

A *continued fraction* is a finite formal expression of the form

$$a_1 + \cfrac{1}{a_2 + \cfrac{1}{a_3 + \cfrac{1}{\ddots + \cfrac{1}{a_n}}}}$$

where all the a_i are in $\mathbb{Z}$ and $a_n \neq 0$. We shall denote it by $C(a_1, a_2, \ldots, a_n)$.

Clearly, each continued fraction can be evaluated to give a rational number. Conversely, every element of $\mathbb{Q}$ can be expanded as a continued fraction. This is an easy corollary of the Euclidean algorithm.

Euclid's division algorithm is written below. Its inputs are coprime integers p and q.

$r_{-1} := p$
$r_0 := q$
$i := -1$
repeat
    $i := i + 1$
    find a_{i+1} and r_{i+1} such that $r_{i-1} = a_{i+1} r_i + r_{i+1}$ and
                $0 \leqslant r_{i+1} < r_i$
    print a_{i+1}
until $(r_{i+1} = 0)$

The output is a sequence of integers $a_1, a_2, \ldots, a_n$ which correspond to the numbers in a continued fraction expansion for p/q. This is easily seen with an example. Each iteration of the 'repeat–until' loop expresses the integer r_{i-1} as a product plus a remainder. The numbers printed out are shown in bold.

$$
\begin{aligned}
159 &= \mathbf{3} \cdot 46 + 21, \\
46 &= \mathbf{2} \cdot 21 + 4, \\
21 &= \mathbf{5} \cdot 4 + 1, \\
4 &= \mathbf{4} \cdot 1 + 0.
\end{aligned}
$$

This calculation can be rewritten to make its relationship to continued fractions more obvious:

$$
\begin{aligned}
\tfrac{159}{46} &= 3 + \tfrac{21}{46} = 3 + \tfrac{1}{46/21}, \\
\tfrac{46}{21} &= 2 + \tfrac{4}{21} = 2 + \tfrac{1}{21/4}, \\
\tfrac{21}{4} &= 5 + \tfrac{1}{4};
\end{aligned}
$$

hence
$$
\frac{159}{46} = 3 + \cfrac{1}{2 + \cfrac{1}{5 + \cfrac{1}{4}}} = C(3, 2, 5, 4).
$$

A continued fraction expansion is not unique. We can change the second part of the rule for choosing a_{i+1} and r_{i+1}. We could, for example, try the condition 'a_{i+1} is even' as shown below.

$$
\begin{aligned}
\tfrac{159}{46} &= 4 - \tfrac{25}{46} = 4 - \tfrac{1}{46/25}, \\
\tfrac{46}{25} &= 2 - \tfrac{4}{25} = 2 - \tfrac{1}{25/4}, \\
\tfrac{25}{4} &= 6 + \tfrac{1}{4};
\end{aligned}
$$

hence
$$
\frac{159}{46} = 4 + \cfrac{1}{-2 + \cfrac{1}{6 + \cfrac{1}{4}}} = C(4, -2, 6, 4).
$$

However, care needs to be taken in choosing a suitable rule. With Euclid's rule '$0 \leqslant r_{i+1} < r_i$' the algorithm is guaranteed to terminate because the sequence of r_i's is strictly decreasing. However, if both p and q are odd, the 'a_{i+1} is even' algorithm will not terminate.

The following formula, due to Lagrange, can be used to manipulate continued fractions:

$$x + \cfrac{1}{-y} \quad = \quad (x-1) + \cfrac{1}{1 + \cfrac{1}{(y-1)}}.$$

Careful application of this formula can reduce the number of negative coefficients in the expansion. A continued fraction is called *regular* if all the coefficients, with the possible exception of a_1, are positive. By repeated use of Lagrange's formula, any continued fraction can be converted into a regular one. Regular continued fractions are almost canonical:

Theorem 8.6.1.   *If $C(a_1, a_2, \ldots, a_n)$ and $C(b_1, b_2, \ldots, b_m)$ are regular continued fractions that evaluate to the same rational number and $n \leqslant m$ then*

(a) *$m = n$ and $b_i = a_i$ for all i, or*

(b) *$m = n + 1$, $b_i = a_i$ for all $i < n$, $b_n = a_n - 1$ and $b_m = 1$.*

The purpose of introducing continued fractions is the following theorem.

Theorem 8.6.2 (Conway).   *Two rational tangles constructed from integer sequences are isotopic if and only if their associated continued fractions evaluate to the same rational number.*

PROOF. Let us recast Conway's algorithm in terms of matrices. As before, its input is a sequence of integers $a_1, a_2, \ldots, a_n$. The starting tangle t_∞ is replaced by a column vector, and each iteration of the algorithm is now a matrix–vector multiplication, rather than an operation of twisting or reflecting a tangle.

$$t := \begin{pmatrix} 1 \\ 0 \end{pmatrix}$$

for each i from n down to 1 do
{
      $t := F t$
      $t := H^{a_i} t$
}

As before, we define the matrix $M(n)$ to be

$$M(n) = H^n F = \begin{pmatrix} n & 1 \\ 1 & 0 \end{pmatrix}.$$

Multiplication by M exactly matches the induction step in the Euclidean algorithm:

$$M(a_{i+1}) \begin{pmatrix} r_i \\ r_{i+1} \end{pmatrix} = \begin{pmatrix} a_{i+1} r_i + r_{i+1} \\ r_i \end{pmatrix} = \begin{pmatrix} r_{i-1} \\ r_i \end{pmatrix}.$$

When the Euclidean algorithm terminates we have $r_n = 0$ and $r_{n-1} = 1$, so the whole division process can be summarised as

$$\begin{pmatrix} p \\ q \end{pmatrix} = M(a_1) M(a_2) \ldots M(a_{n-1}) M(a_n) \begin{pmatrix} 1 \\ 0 \end{pmatrix}.$$

This is exactly what the matrix–vector form of Conway's algorithm calculates. □

8.7 Rational links

The closure (numerator or denominator) of a rational tangle is a rational knot or 2-component link. Theorem 4.10.3 shows that the converse is also true: a rational knot or link is the closure of some rational tangle in pillowcase form. We have also seen that Conway's procedure generates all rational tangles. This has the following useful consequence.

Theorem 8.7.1. *A rational link has an alternating diagram.*

PROOF. Suppose that a rational link L is the numerator of rational tangle $t_{p/q}$. There is a continued fraction expansion of p/q in which all the coefficients have the same sign. The corresponding tangle diagram is alternating. □

The classification of rational tangles does not immediately give us a classification of rational links. When we close a rational tangle to form a link, L, the isotopies we can perform are no longer restricted to the arcs of $L \cap B$, nor do we have to keep the points of $L \cap \partial B$ fixed. This additional freedom means that some rational tangles close to give the same link. For example, the numerators of the tangles corresponding to the fractions $3/1 = C(3)$ and $-3/2 = C(-2, 2)$ are both left-handed trefoils.

The classification of rational links uses a similar technique to the classification of rational tangles, except that now everything takes place in three dimensions. Our classification of rational tangles involved looking at the 2-fold covering of S^2 with four branch points. For a rational link we consider the 2-fold covering of S^3 with the link as a branch set. The resulting covering space is a 3-manifold known as a lens space. The classification of lens spaces [154] leads to the following theorem [163].

Theorem 8.7.2. *The numerators of two rational tangles p_1/q_1 and p_2/q_2 are isotopic if and only if $p_1 = p_2$ and either $q_1 \equiv q_2$ mod p_1 or $q_1 q_2 \equiv 1$ mod p_1.*

Let us use the case $p = 7$ as an example. The following table lists the elements of $\mathbb{Q}$ larger than 1 with numerator 7, their continued fraction expansions arising from the standard Euclidean algorithm, and the associated rational knots.

$7/1$	$C(7)$	7_1
$7/2$	$C(3, 2)$	5_2^*
$7/3$	$C(2, 3)$	5_2
$7/4$	$C(1, 1, 3)$	5_2^*
$7/5$	$C(1, 2, 2)$	5_2
$7/6$	$C(1, 6)$	7_1^*

We see that fractions $7/2$ and $7/4$ give rise to the same knot, thus satisfying the theorem: $2 \times 4 \equiv 1$ mod 7. Similarly with fractions $7/3$ and $7/5$ since $3 \times 5 \equiv 1$ mod 7.

We know that the rational tangle corresponding to the fraction $-p/q$ is the mirror image of the tangle for p/q. Since we are working modulo p, the fractions p/q and $\frac{p}{q-p}$ are equivalent. This explains the following pattern exhibited in the table: the knots with fractions p/q and $\frac{p}{p-q}$ are mirror images.

There is a choice of normal form for the fraction of a rational link. Because we need only consider q modulo p we can assume that $0 < q < p$ as in the examples just listed. Another frequently used normalisation is to choose $-p < q < p$ and q odd. This is possible because if q is even p cannot be even (since p and q are coprime) and we can replace q with $q - p$ without changing the link type.

We have already seen that the standard Euclidean algorithm produces a canonical form of continued fraction in which all the coefficients have the same sign, and that this has the consequence that rational links are alternating. By changing the rule for choosing the coefficients, we can produce different canonical forms. One of the most useful from the knot theory viewpoint is the one where all the coefficients are even.

Corollary 8.7.3. *A rational link has a continued fraction description of the form $C(2a_1, 2a_2, \ldots, 2a_{n-1}, 2a_n)$.*

PROOF. If both p and q are odd, we can replace q with $q - p$ to obtain an isotopic link with a fraction whose denominator is even. It is then possible to derive a continued fraction in which all the coefficients are even (Exercise 8.10.8). □

The fact that all the coefficients are even means that the orientations of the strings connecting the twist-boxes in a standard diagram of a rational link are known: they are independent of the coefficients. This means it is straightforward to apply Seifert's algorithm or skein relations.

Theorem 8.7.4. *The Conway polynomial of the rational link with continued fraction $C(2a_1, 2a_2, \ldots, 2a_{n-1}, 2a_n)$ is*

$$\begin{pmatrix} 1 & 0 \end{pmatrix} \begin{pmatrix} (-1)^{n-1}a_n z & 1 \\ 1 & 0 \end{pmatrix} \cdots \begin{pmatrix} -a_2 z & 1 \\ 1 & 0 \end{pmatrix} \begin{pmatrix} a_1 z & 1 \\ 1 & 0 \end{pmatrix} \begin{pmatrix} 1 \\ 0 \end{pmatrix}.$$

PROOF. Let $\nabla_i(z)$ denote the Conway polynomial of the rational link formed from the first i coefficients:

$$\nabla_i(z) = \nabla\Big(C(2a_1, 2a_2, \ldots, 2a_{i-1}, 2a_i)\Big).$$

We extend the notation setting $\nabla_0(z) = 1$ and $\nabla_{-1}(z) = 0$, these corresponding to the trivial knot and the trivial 2-component link, respectively.

We want to apply the Conway skein relation to the left-most twist-box, the one containing a_n full twists (see Figure 3.9). If n is odd then the sign of the crossings matches the sign of a_n; if n is even then the crossings have the opposite sign to a_n. Applying the skein relation a_n times we get

$$\nabla_n(z) = \nabla_{n-2}(z) + (-1)^{n-1} a_n z \nabla_{n-1}(z).$$

This recurrence relation can be expressed in matrix form:

$$\begin{pmatrix} \nabla_n(z) \\ \nabla_{n-1}(z) \end{pmatrix} = \begin{pmatrix} (-1)^{n-1}a_n z & 1 \\ 1 & 0 \end{pmatrix} \begin{pmatrix} \nabla_{n-1}(z) \\ \nabla_{n-2}(z) \end{pmatrix}.$$

Applying this relationship recursively gives the sequence of matrices above; the process terminates with $\nabla_0(z)$ and $\nabla_{-1}(z)$ in the right-hand vector. □

The Conway polynomial does not distinguish between all rational links. On page 158, we listed five pairs of knots with the same Alexander polynomial. In two of these cases, both knots in the pair are rational knots:

$$
\begin{array}{lll}
7_4 & {}^{15}\!/_4 & C(3,\,1,\,3) \\
9_2 & {}^{15}\!/_7 & C(2,\,7)
\end{array}
\left.\rule{0pt}{24pt}\right\} \quad \nabla(z) = 1 + 4z^2;
$$

$$
\begin{array}{lll}
8_{14} & {}^{31}\!/_{12} & C(2,\,1,\,1,\,2,\,2) \\
9_8 & {}^{31}\!/_{11} & C(2,\,1,\,4,\,2)
\end{array}
\left.\rule{0pt}{24pt}\right\} \quad \nabla(z) = 1 - 2z^4.
$$

Corollary 8.7.5. *Let L be the rational link with continued fraction $C(2a_1, 2a_2, \ldots, 2a_{n-1}, 2a_n)$. Its genus is given by*

$$
g(L) \;=\; \left\{
\begin{array}{ll}
{}^1\!/_2\, n & \text{if } n \text{ is even,} \\
{}^1\!/_2\,(n-1) & \text{if } n \text{ is odd.}
\end{array}
\right.
$$

PROOF. If all the coefficients in the continued fraction are even, the tangle numerator is a knot when n is even and a 2-component link when n is odd. Therefore, $n \bmod 2 = \mu(L) - 1$.

It is clear from the previous theorem that $\nabla_L(z)$ has degree n. Furthermore, a rational link is alternating (Theorem 8.7.1). Hence, by Corollary 7.6.5,

$$
2g(L) \;=\; \operatorname{maxdeg} \nabla_L(z) - \mu(L) + 1 \;=\; n - (n \bmod 2).
$$

□

The projection surface constructed from the 'all even coefficients' continued fraction diagram has minimal genus (Exercise 8.10.9).

We can find the values of other invariants of rational links directly from the fraction ${}^p\!/_q$ or one of its continued fraction expansions. As a second corollary to Theorem 8.7.4 we shall derive the determinant. First we need the following lemma.

Lemma 8.7.6. *Suppose the continued fraction $C(a_1, a_2, \ldots, a_{n-1}, a_n)$ evaluates to ${}^p\!/_q$, and let $i = \sqrt{-1}$. Then*

$$
\begin{pmatrix} (-1)^{n+1}\,i\,a_n & 1 \\ 1 & 0 \end{pmatrix} \cdots \begin{pmatrix} -i\,a_2 & 1 \\ 1 & 0 \end{pmatrix} \begin{pmatrix} i\,a_1 & 1 \\ 1 & 0 \end{pmatrix} \begin{pmatrix} 1 \\ 0 \end{pmatrix}
$$

evaluates to $\begin{pmatrix} i\,p \\ q \end{pmatrix}$ *when n is odd, and* $\begin{pmatrix} p \\ i\,q \end{pmatrix}$ *when n is even.*

PROOF. Using the relation $r_{j-1} = a_{j+1}\,r_j + r_{j+1}$, the two following observations show that this procedure is just another (slightly obscure) encoding of the Euclidean algorithm:

$$\begin{pmatrix} i\,a_{j+1} & 1 \\ 1 & 0 \end{pmatrix} \begin{pmatrix} r_j \\ i\,r_{j+1} \end{pmatrix} = \begin{pmatrix} i\,a_{j+1}\,r_j + i\,r_{j+1} \\ r_j \end{pmatrix} = \begin{pmatrix} i\,r_{j-1} \\ r_j \end{pmatrix};$$

$$\begin{pmatrix} -i\,a_{j+1} & 1 \\ 1 & 0 \end{pmatrix} \begin{pmatrix} i\,r_j \\ r_{j+1} \end{pmatrix} = \begin{pmatrix} -i^2\,a_{j+1}\,r_j + r_{j+1} \\ i\,r_j \end{pmatrix} = \begin{pmatrix} r_{j-1} \\ i\,r_j \end{pmatrix}.$$

$\square$

Theorem 8.7.7. *A rational link with fraction p/q has determinant p.*

PROOF. First express p/q as a continued fraction with even coefficients: $C(2a_1, 2a_2, \ldots, 2a_{n-1}, 2a_n)$. Combining Theorem 7.1.4(a), which states that $|\Delta_L(x)| = \det(L)$ when $x = \pm i$, and the Conway polynomial substitution $z = x^{-1} - x$, we obtain $\det(K) = |\nabla_K(2i)|$. Substituting $z = 2i$ into the matrix expression for ∇_K we get

$$\begin{pmatrix} 1 & 0 \end{pmatrix} \begin{pmatrix} (-1)^{n+1}\,i\,2\,a_n & 1 \\ 1 & 0 \end{pmatrix} \cdots \begin{pmatrix} -i\,2\,a_2 & 1 \\ 1 & 0 \end{pmatrix} \begin{pmatrix} i\,2\,a_1 & 1 \\ 1 & 0 \end{pmatrix} \begin{pmatrix} 1 \\ 0 \end{pmatrix}.$$

By the preceding lemma, this evaluates to p or $i\,p$. Thus $|\nabla_K(2i)| = p$. $\square$

Corollary 8.7.8. *A rational link with fraction p/q is a knot if p is odd, and a link if p is even.*

PROOF. A rational link has one or two components (Exercise 4.11.15). A knot has odd determinant; a link has even determinant (see Lemma 6.7.6 and Theorem 6.7.1). $\square$

8.8 Enumerating rational links

Conway's motivation for studying tangles was to extend the catalogues. In the next section, we shall see how rational tangles can be used to enumerate links in general, but here we shall concentrate on finding the first few rational links. The problem is reduced to listing sequences of integers, and noting which patterns lead to isotopic links. This technique is so powerful that

Conway claims to have verified the Tait–Little tables 'in an afternoon'. He then went on to list the 11-crossing knots and 10-crossing links.

Before starting the enumeration, we need to know which sequences give rise to equivalent links.

Theorem 8.8.1.   *Rational links have the following symmetries.*

(a) *The links with the following continued fractions are isotopic:*

$$C(a_1, a_2, \ldots, a_n), \qquad C(a_1, a_2, \ldots, a_n \pm 1, \mp 1).$$

(b) *If $\varepsilon = (-1)^{n-1}$ the links with the following continued fractions are isotopic:*

$$C(a_n, a_{n-1}, \ldots, a_2, a_1), \qquad C(\varepsilon\, a_1, \varepsilon\, a_2, \ldots, \varepsilon\, a_{n-1}, \varepsilon\, a_n).$$

(c) *The links with the following continued fractions are mirror images:*

$$C(a_1, a_2, \ldots, a_{n-1}, a_n), \qquad C(-a_1, -a_2, \ldots, -a_{n-1}, -a_n).$$

PROOF. Exercise 8.10.3.   □

Using these three symmetries, we can enumerate the integer sequences that correspond to rational knots and links.

A rational link (or its mirror image) has a regular continued fraction expansion in which all the integers are positive. By Theorem 8.8.1(a) we can discard all sequences that end in a 1, and this makes the regular sequence unique. By part (b) we do not need to keep both a sequence and its reverse. Applying these simple rules to the partitions of the first four integers, we see that we need only keep the sequences shown in bold:

$$\mathbf{1}, \mathbf{2}, 11, \mathbf{3}, 21, 12, 111, \mathbf{4}, 31, \mathbf{22}, 13, 211, 121, 112, 1111.$$

These sequences correspond in order to the trivial knot, the Hopf link, the trefoil, the (2,4) torus link, and the figure-8 knot.

Continuing in this fashion, we find that for knots and links with up to seven crossings, the sequences for rational knots are

$$3, 22, 5, 32, 42, 312, 2112, 7, 52, 43, 322, 313, 2212, 21112$$

and the sequences for rational 2-component links are

$$2, 4, 212, 6, 33, 222, 412, 232, 3112.$$

By combining parts (b) and (c) of Theorem 8.8.1 we see that a sequence represents an amphicheiral knot or link if and only if the sequence is palindromic (equal to its reverse) and of even length (n even). This shows that the only amphicheiral knots in the list are the figure-8 knot (sequence 22) and the knot 6_3 (sequence 2112); all of the links are cheiral.

The set of knots is actually complete: there are no other knots with fewer than eight crossings. The pattern does not continue any further: only 12 of the 21 8-crossing knots, and 21 of the 49 9-crossing knots are rational. For the 2-component links, the set contains all those with at most six crossings but only three of the eight 7-crossing links are rational. The technique will never generate links with more than two components.

8.9   Applications of rational tangles

To close this chapter we mention three applications of rational tangles. The first two are related to the enumeration of links; the third is in the field of molecular biology.

First we conclude the story of Conway's method of enumerating links. As simply closing rational tangles does not produce the full range of links, we must explore how to glue the tangles together in various ways. One way to present the gluing information is to take a 4-valent plane graph and then replace each vertex with a tangle. If we restrict ourselves to plugging in rational tangles then the set of graphs we need to consider is quite large. However, notice that if two tangles t_A and t_B are connected by a 2-gon in the graph then we can collapse the 2-gon, removing it from the graph, and place the partial sum $t_A \oplus t_B$ in the coalesced vertex. A tangle that can be produced by repeatedly taking partial sums of rational tangles is called an *algebraic* tangle. By plugging algebraic tangles into a graph, we can ignore graphs that contain 2-gons. This leads to a substantial reduction in the number of graphs to be considered. In fact for up to eight vertices the only possibilities are shown in Figure 8.8. The first graph produces the rational links (as we have seen) and also the pretzel knots: $P(p, q, r)$ is the numerator of $t_{1/p} \oplus t_{1/q} \oplus t_{1/r}$. The most general type of link that can be formed from the closure of an algebraic tangle is sometimes called an *arborescent* link because the 2-gons in the graph form a tree-like structure. The other 4-valent graphs with no 2-gons are called *polyhedra*; they have at least six vertices. After being replaced by a tangle, each vertex must contribute at least one crossing to the diagram. So to enumerate links with small crossing number, the substituted tangles cannot be very complicated.

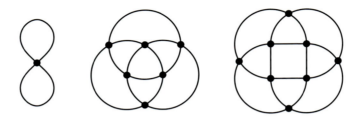

Figure 8.8. The first three 4-valent graphs with no 2-gons.

Of the 84 non-trivial knots up to nine crossings, 50 are rational, a further 21 are arborescent, and only 13 come from polyhedral graphs.

The task of enumerating knots prompts the following two questions:

1. How many prime knots are there with crossing number n?

2. How does this number grow as a function of n?

The answers to question 1 for $n \leqslant 16$ are listed in Figure 4.10. The second question is more difficult because we do not know how to prove that a given diagram has minimal crossing number. In the next chapter we shall see that irreducible alternating diagrams are always minimal (Theorem 9.5.4 and Corollary 9.5.10). The method of enumerating rational links by listing sequences of integers produces such diagrams. By counting the number of partitions of n, and taking various symmetry properties into account, Claus Ernst and Dewitt Sumners calculated that the number of rational knots of order n is at least $1/3 \left(2^{n-2} - 1 \right)$ when both knots in a cheiral pair are included in the count [156]. As rational knots are known to be prime (Exercise 4.11.19) this gives a lower bound on the growth rate. A more accurate bound has been found by Carl Sundberg and Morwen Thistlethwaite [168].

The third application concerns the action of enzymes which alter the topology of DNA molecules. Such modifications are necessary during the processes of replication and recombination and include changing the level of supercoiling, passing one strand of DNA through a temporary gap in another, and breaking a pair of strands apart and recombining them. The last two actions are similar to the skein operations in which D_+ is replaced by D_- or D_0 respectively.

During the reaction, the *substrate* (input) molecule binds to the enzyme forming a *complex*, the enzyme performs its action, and releases the *product*. It is impossible to observe directly how the enzyme executes its function:

electron micrographs of the DNA–enzyme complex show a blob with strands coming out but not the internal arrangement. Indirect evidence can be obtained by placing synthetic unknotted circular DNA in the presence of an enzyme and examining the products. Tangle theory can then be used to deduce the configuration of the strands within the complex and the action of the enzyme.

The mathematical model of the situation assumes that

- the DNA–enzyme complex can be modelled as a 2-tangle with the enzyme forming the ball and two strands of DNA being the arcs,
- each arc within the tangle is unknotted,
- the enzyme replaces one tangle by another.

We also make the following biological assumption:

- the action performed by the enzyme is constant, is independent of the shape and topological type of the substrate, and does not affect parts of the molecule lying outside the enzyme.

As the substrate molecule approaches and binds to the enzyme, it is likely that each strand lies on the outside of the enzyme-ball. This explains why the tangle strings are assumed to be unknotted. If both strands were to sit on the surface simultaneously, the tangle would be rational.

The results from one experiment are listed below. The enzyme used was Tn3 resolvase. It acts *processively*, meaning that it can perform its function more than once in the same place before releasing the molecule. The substrate molecule is modelled as $N(t_S \oplus t_A)$ where the tangle t_A is assumed to be inside the enzyme. The enzyme replaces t_A with one or more copies of t_R. The amount of product produced decreases with the complexity of the product. After detecting the first three products listed below, the tangles involved could be calculated by solving simultaneous equations. The model predicted the knot 6_2 as a further product and this was later detected. See any of [157, 165, 166, 167] for more details.

Substrate	$N(t_S \oplus t_A)$	trivial knot,
Product 1	$N(t_S \oplus t_R)$	Hopf link,
Product 2	$N(t_S \oplus t_R \oplus t_R)$	figure-8 knot,
Product 3	$N(t_S \oplus t_R \oplus t_R \oplus t_R)$	Whitehead link,
Product 4	$N(t_S \oplus t_R \oplus t_R \oplus t_R \oplus t_R)$	6_2.

8.10 Exercises

1. Show that the continued fractions $C(4, 1, 5)$ and $C(4, 2, -2, 2, -2, 2)$ evaluate to the same rational. (These tangles close to give the two diagrams in Figure 3.15.) Draw a sequence of pictures to show the tangles are isotopic.

2. Verify that applying Lagrange's formula does not change the value of a continued fraction. Draw pictures to show the corresponding isotopy of rational tangles.

3. Prove Theorem 8.8.1.

4. Show that a rational link with fraction p/q is amphicheiral if and only if $q^2 \equiv -1 \bmod p$.

5. Enumerate the 12 rational knots and 6 rational links with 8 crossings. Are any of them amphicheiral?

6. Suppose that t_A, t_B and their partial sum $t_A \oplus t_B$ are all rational tangles. Find a relationship between their continued fractions.

7. Draw pictures to verify the following relationships topologically: $R = HV^{-1}H$, $V^{-1} = RHR^{-1}$, $V = HRH$, $R^2 = I$, $(HR)^3 = I$.

8. Find a variant of the Euclidean algorithm that produces continued fractions with even coefficients when the inputs are coprime integers p and q, not both odd. Prove the algorithm always terminates. Is the result unique?

9. Show that the projection surface constructed from a diagram coming from a continued fraction in which all the coefficients are even has minimal genus.

10. Prove that $\mathrm{GL}(2, \mathbb{Z})$ is generated by H, V and F (see page 199 for definitions of these matrices).

11. Prove that $\mathrm{PSL}(2, \mathbb{Z})$ is generated by any pair of H, V and R.

9 More Polynomials

The story we have seen so far shows the state of combinatorial knot theory in the early 1980s. There were various geometrically motivated numerical invariants, some dating from the birth of the subject, all of which were difficult to calculate; there were a few computable invariants derived from Seifert matrices and skein theory; and there were some relationships between the two types.

In this chapter we shall introduce some of the new polynomial-valued link invariants that resulted from a chance discovery linking knot theory with theoretical physics. We shall concentrate on the combinatorial properties of Louis Kauffman's bracket polynomial and its relationship to crossing number.

First, however, we shall tell the story of the event that sparked a revolution.

9.1 Discovery of the Jones polynomial

The technical details in this section are included only to give an idea of what was involved in the discovery — they are not used in the rest of the book.

In §7.10 and Exercise 7.12.18 we introduced the n-string braid group $\mathbb{B}_n$. We also noted that a function which takes the same value on braids that close to produce the same link is a link invariant. However, the task of constructing such functions is not easy.

The first problem is to determine which braids close to a given link. Andrei Markov[*] proposed the following set of moves [195]: the analogous Reidemeister moves are indicated in parentheses.

---

[*]Son of Andrei Markov the probability theorist, who gave us Markov chains and Markov processes.

- Braid relations:
  $$\sigma_i \sigma_j = \sigma_j \sigma_i \text{ when } |i - j| \geqslant 2 \qquad \text{(isotopy in the plane)},$$
  $$\sigma_i \sigma_i^{-1} = 1 = \sigma_i^{-1} \sigma_i \qquad \text{(Reidemeister 2)},$$
  $$\sigma_i \sigma_{i+1} \sigma_i = \sigma_{i+1} \sigma_i \sigma_{i+1} \qquad \text{(Reidemeister 3)}.$$

- Conjugacy:
  $$\sigma_i \beta \sigma_i^{-1} = \beta = \sigma_i^{-1} \beta \sigma_i.$$

- Stabilisation:
  replace $\beta \in \mathbb{B}_n$ with $\beta \sigma_n^{\pm 1} \in \mathbb{B}_{n+1}$ or vice versa $\qquad$ (Reidemeister 1).

Conjugating a braid by a generator is just a type 2 Reidemeister move on the associated closed braid; the Markov equivalent of the type 1 Reidemeister move changes the number of strings in the braid.

Showing that any two braids which close to the same link are related by a finite sequence of these moves is not a simple translation of the Reidemeister moves: braids are oriented and the Markov moves preserve the string orientations. Proofs that the Markov moves suffice can be found in [4, 170, 196, 242].

The next problem is to map the braid group into another algebraic structure that is better understood than the braid group, and then find some property of this structure that is preserved when the braid is changed by a Markov move. One solution to this problem was described in 1936 by Werner Burau [174]. He mapped the braids onto matrices as follows. Let $\mathrm{GL}(n, \mathbb{Z}[t^{\pm 1}])$ be the group of invertible $n \times n$ matrices whose entries are Laurent polynomials in t with integer coefficients. There is a map from $\mathbb{B}_n$ into $\mathrm{GL}(n, \mathbb{Z}[t^{\pm 1}])$ defined by sending σ_i to the matrix formed by taking the identity and replacing the 2×2 submatrix whose top-left corner is at position (i, i) with

$$\begin{pmatrix} 1 - t & t \\ 1 & 0 \end{pmatrix}.$$

This map is a homomorphism (Exercise 9.8.1) and is called the Burau representation of the braid group. The matrix trace (sum of the diagonal entries) is unchanged by conjugation, and further analysis leads to a function that is also invariant under stabilisation. The resulting link invariant is the Alexander polynomial. This was the only known representation of $\mathbb{B}_n$ for almost 50 years.

In the summer of 1982, Vaughan Jones was presenting a lecture on von Neumann algebras in Geneva. At the end of his talk Didier Hatt-Arnold, a graduate student in the audience, suggested that the relations in his algebraic structures were similar to those in the braid groups. Soon afterwards, Jones worked out how to construct representations of the braid groups into his algebras, but he did not immediately recognise their significance. The following summer, Jones realised that the image of $\mathbb{B}_5$ under one of the representations was the projective symplectic group $\mathrm{PSp}(4, \mathbb{Z}_3)$, the finite simple group of order 25 920. Thinking that this might be of some interest, Jones arranged to discuss his representations with Joan Birman.

Jones' algebras are finite-dimensional algebras of operators on a Hilbert space. Each algebra $A_n(t)$ is a direct sum of matrix algebras, parametrised in some fashion by a real number t. For each admissible value of t, it is possible to construct a representation $\mathbb{B}_n \longrightarrow A_n(t)$ of a braid group. Jones' algebras also support a trace function and this gives rise to a map $\mathbb{B}_n \longrightarrow \mathbb{C}[t^{-1/2}, t^{1/2}]$. The polynomial value produced is not affected when a braid is changed by a Markov move so the map is a link invariant.

Jones travelled to New York in May 1984. He and Birman soon showed that this was not just another technique for deriving the Alexander polynomial. One simple test proved that this invariant was new: it could distinguish the left-handed and right-handed trefoils! Jones later established that his polynomial also satisfies a skein relation:

$$t\, V(L_+) - t^{-1}\, V(L_-) \;=\; (t^{1/2} - t^{-1/2})\, V(L_0).$$

This discovery had a tremendous impact, and not only on knot theory. Once it was known that the Alexander polynomial was not the only polynomial link invariant, people started to search for more — some using combinatorics and others following the algebraic route used by Jones. Close connections with physics generated a lot of interdisciplinary research, and polynomials were defined via physical methods related to statistical mechanics [189], where the Yang–Baxter equation provides an analogue of the third braid relation [209], and quantum groups. These new link invariants were also extended to give invariants of 3-manifolds [191, 203, 210].

In this chapter we shall study one of these new invariants in detail; we shall see applications of two more in Chapter 10.

9.2 The Kauffman bracket

Definition 9.2.1. The *bracket polynomial* of an unoriented diagram D, denoted by $\langle D \rangle$, is a Laurent polynomial in a single variable A defined by the three following axioms.

1. Normalisation: $\langle \bigcirc \rangle = 1$ where $\bigcirc$ denotes the diagram of one component and no crossings.

2. Delta: $\langle D \sqcup \bigcirc \rangle = \delta \langle D \rangle$ where $\delta = -A^{-2} - A^2$ and $D \sqcup \bigcirc$ denotes the diagram D together with a single component that does not cross itself or D.

3. Skein relation: $\langle D_+ \rangle = A \langle D_0 \rangle + A^{-1} \langle D_\infty \rangle$ where diagrams D_+, D_0 and D_∞ differ as shown in Figure 9.1.

There are several things about this definition that should be noted.

- This polynomial is defined on diagrams and not on links: the axiom of invariance under ambient isotopy that we used in Definition 7.4.1 of the Conway polynomial is absent.

- The three oriented diagrams of Figure 6.8 are defined in an absolute manner. However, the unoriented diagrams used here are only defined in relation to each other: there is no way to determine whether a given crossing in a diagram is of type D_+ or D_-, but once a choice has been made, the other three diagrams are uniquely determined.

- Combining the previous comment with the skein relation gives us

$$\langle D_- \rangle \;=\; A \langle D_\infty \rangle + A^{-1} \langle D_0 \rangle.$$

- In the second axiom, there is no restriction on where in the plane the trivial component $\bigcirc$ is placed: it can be in any region of the plane; it can surround or separate other components of the diagram.

The axioms suffice for a recursive computation of $\langle D \rangle$: by applying the skein relation repeatedly, a diagram can be transformed into a set of diagrams with no crossings, and the value on each of these can be calculated from the first two axioms: the polynomial of n disjoint circles in the plane is δ^{n-1}.

If the bracket polynomial is to be a link invariant then its value should be equal on diagrams that differ by sequences of Reidemeister moves. We shall consider them in reverse order.

Figure 9.2 shows applications of the bracket skein relation to the before and after situations of a type 3 move. It is clear that the first and third pictures in the bottom row differ by two applications of a type 2 move. So, if $\langle \ \rangle$ is invariant under type 2 moves, it is also invariant under type 3 moves.

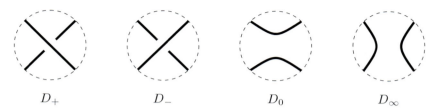

$$D_+ \qquad\qquad D_- \qquad\qquad D_0 \qquad\qquad D_\infty$$

Figure 9.1. Four unoriented diagrams which differ only inside a small neighbourhood.

So let us consider type 2 moves. Figure 9.3 shows the effects of applying the bracket skein relation to both crossings involved in a type 2 move. In order for the evaluation of $\langle\,\rangle$ to be the same before and after the move we must have

$$(A^2 + \delta + A^{-2})\,D_0 + D_\infty \;=\; D_\infty.$$

This only happens if the coefficient of D_0 is zero, which forces $\delta = -A^{-2} - A^2$ (its value in the axioms). Hence $\langle\,\rangle$ is invariant under type 2 and type 3 moves.

Figure 9.4 shows the application of the skein relation to a type 1 move involving a negative curl (the sign of the *oriented* crossing is -1). For this move to leave $\langle\,\rangle$ invariant, we must have

$$(A + \delta\,A^{-1}) \;=\; -A^{-3} \;=\; 1.$$

Similarly, starting with a positive curl we can deduce $-A^3 = 1$. Unfortunately, satisfying these requirements forces us to set $A = \sqrt[3]{-1} \in \mathbb{C}$ which means $\langle\,\rangle$ is no longer a polynomial.

As we shall see, an invariant of Reidemeister moves 2 and 3 is quite a useful tool. Furthermore, the behaviour of $\langle\,\rangle$ under type 1 moves is so well

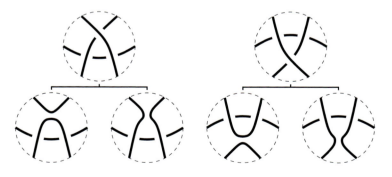

Figure 9.2. Behaviour of the bracket under type 3 Reidemeister moves.

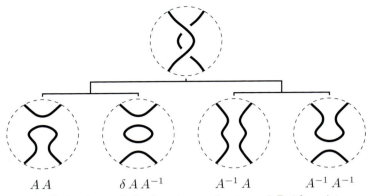

$$A\,A \qquad\qquad \delta\,A\,A^{-1} \qquad\qquad A^{-1}\,A \qquad\qquad A^{-1}\,A^{-1}$$

Figure 9.3. Behaviour of the bracket under type 2 Reidemeister moves.

controlled that it is possible to use it as the basis of a polynomial invariant of oriented links. To complete the construction, we use that fact that the writhe of an oriented diagram is also invariant under moves 2 and 3 and behaves in a predictable way under move 1.

By analogy with the arithmetic notation whereby $|x|$ denotes the number x with its sign ignored, we use $|L|$ to denote the link L with its orientation ignored. Similarly $|D|$ is an unoriented diagram.

Definition 9.2.2. The *normalised bracket polynomial* of an oriented diagram D is

$$\widetilde{V}_D(A) \;=\; (-A^{-3})^{w(D)} \,\langle\,|D|\,\rangle(A).$$

It is invariant under all three Reidemeister moves and hence is a well-defined invariant of oriented links. Thus we can write $\widetilde{V}_L$ without reference to a specific diagram of the link L.

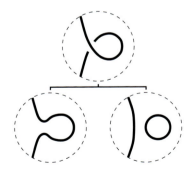

Figure 9.4. Behaviour of the bracket under type 1 Reidemeister moves.

Before we continue, take a moment to reflect on the simplicity of this scheme and the straightforward method used to establish its invariance in just a few pages. Compare it with the amount of work that went into defining the Alexander polynomial: Seifert's algorithm to obtain a spanning surface, homology theory to define the Seifert matrix, then proving invariance of the polynomial under surgery on the surface. The bracket construction is also far simpler than the algebraic process outlined above that led to the Jones polynomial, although, as we shall see, the normalised bracket and Jones polynomials are equivalent under a change of variable.

Despite its simple definition, this new polynomial invariant has power. Applying the link invariants described in earlier chapters to the prime knots up to order 9 left five pairs unresolved (see page 158). Of these, two pairs are rational knots and can be shown to be distinct by their continued fraction expansions (see page 208). The normalised bracket polynomial is able to distinguish all five pairs: this completes the proof that all the knots listed in the catalogue up to order 9 are distinct. The polynomial $\widetilde{V}$ also distinguishes the problem knots of §7.11 (8_{20} and $3_1 \# 3_1^*$), completing the classification of the knots in Figure 1.6.

Notwithstanding these successes, our experience with other invariants leads us to surmise that the new invariant will be unable to distinguish some knots. This is indeed the case. The knots 8_9 and $4_1 \# 4_1$ have the same $\widetilde{V}$ but different Alexander polynomials. This example is useful because it shows that the two polynomials really are different invariants: we cannot obtain Δ as a special case of $\widetilde{V}$. Yet even their combined power is not enough to distinguish the knots in the catalogue to order 10: for the knots 8_8^* and 10_{129} (see Figure 7.5) both polynomials are the same.

9.3 Properties of the bracket polynomial

One property that immediately distinguishes $\widetilde{V}$ from the Alexander polynomial is its ability to detect cheirality. Let D^* denote the diagram obtained from D by reflection in the plane of the diagram. This switches all the crossings so

$$\langle D^* \rangle(A) \;=\; \langle D \rangle(A^{-1}) \quad \text{and} \quad \widetilde{V}(D^*)(A) \;=\; \widetilde{V}(D)(A^{-1}).$$

Thus an amphicheiral link has a symmetric polynomial: $\widetilde{V}(A) = \widetilde{V}(A^{-1})$.

We have seen that signature can be used to show some links are cheiral (Corollary 6.7.3) but this test fails for 6_1. The new method shows that 6_1

is indeed cheiral:

$$\widetilde{V}(6_1) \; = \; A^{-16} - A^{-12} + A^{-8} - 2A^{-4} + 2 - A^4 + A^8.$$

On the other hand, 9_{42} has a symmetric bracket polynomial but it is cheiral because its signature is 2.

Theorem 9.3.1. *The normalised bracket polynomial has the following properties.*

(a) *If L is a link then $\widetilde{V}(-L) = \widetilde{V}(L)$.*

(b) *If L is a link then $\widetilde{V}(L^*)(A) = \widetilde{V}(L)(A^{-1})$.*

(c) *If $L_1 \sqcup L_2$ is a split link then $\widetilde{V}(L_1 \sqcup L_2) = \delta \, \widetilde{V}(L_1) \, \widetilde{V}(L_2)$.*

(d) *If a link can be factorised as $L_1 \,\#\, L_2$ then $\widetilde{V}(L_1 \,\#\, L_2) = \widetilde{V}(L_1) \, \widetilde{V}(L_2)$.*

PROOF. Exercise 9.8.3.   □

The first property noted in the preceding theorem is that reversing the orientations of *all* the components of a link does not change the polynomial. When the orientations of *some* of the components are changed, the polynomial may be affected. In the case of the Alexander–Conway polynomial, the effect can be considerable. However, since the normalised bracket is defined largely on unoriented diagrams, and only later adjusted by a multiplier that depends on the orientation, it is altered in a more controlled and predictable manner [192, 197].

Theorem 9.3.2. *Let $L \cup M$ be an oriented link with components $L_1, \ldots, L_p, M_1, \ldots, M_q$. If the orientations of the components in the sublink L are reversed then the polynomial changes as follows:*

$$\widetilde{V}(-L \cup M) \; = \; A^{12\lambda} \, \widetilde{V}(L \cup M) \quad \text{where} \quad \lambda = \sum_{i=1}^{p} \sum_{j=1}^{q} \mathrm{lk}(L_i, M_j).$$

PROOF. Exercise 9.8.4.   □

For the Alexander polynomial there is a formula (Theorem 7.1.4(g)) that relates the polynomials of a satellite link S to those of its companion C and pattern P: $\Delta_S(x) = \Delta_P(x) \Delta_C(x^n)$ where n is the winding number of P. The 12-crossing knot shown in Figure 9.5 has the same $\widetilde{V}$ polynomial as the torus knot 7_1^*. We can build two satellites, each with one of these as a companion, using identical patterns and framing. If a similar satellite

Figure 9.5. This knot (12n749) has the same Alexander and Jones polynomials as 7_1^*. It has signature 2 while $\sigma(7_1) = 6$ so the knots are distinct [171].

formula existed for $\widetilde{V}$ then satellites constructed in this way would have identical polynomials, but simple cables about these two knots disprove this [198]. Note that we cannot use mutant knots as companions in this example as satellites around mutants always share the same polynomial [199].

Finally in this section, we show that Kauffman's combinatorics are equivalent to Jones' algebra.

Theorem 9.3.3. *The normalised bracket polynomial is equivalent to the Jones polynomial under a simple change of variable:*

$$V_L(A^{-4}) = \widetilde{V}_L(A).$$

PROOF. From the bracket skein relation we have

$$\langle D_+ \rangle = A \langle D_0 \rangle + A^{-1} \langle D_\infty \rangle,$$
$$\langle D_- \rangle = A^{-1} \langle D_0 \rangle + A \langle D_\infty \rangle.$$

Multiplying by A and A^{-1} respectively and subtracting gives

$$A \langle D_+ \rangle - A^{-1} \langle D_- \rangle = (A^2 - A^{-2}) \langle D_0 \rangle.$$

These three diagrams can be oriented consistently with each other so we can compare their writhes: $w(D_+) - 1 = w(D_0) = w(D_-) + 1$. The definition of $\widetilde{V}$ can be rearranged to give $\langle D \rangle(A) = (-A^3)^{w(D)} \widetilde{V}_D(A)$. Combining this with the last displayed equation gives

$$(-A^3) A \widetilde{V}(D_+) - (-A^{-3}) A^{-1} \widetilde{V}(D_-) = (A^2 - A^{-2}) \widetilde{V}(D_0)$$

and hence

$$A^4 \, \widetilde{V}(D_+) - A^{-4} \, \widetilde{V}(D_-) \; = \; (A^2 - A^{-2}) \, \widetilde{V}(D_0).$$

This is the skein relation of the Jones polynomial with the substitution given in the statement of the theorem. Both V and $\widetilde{V}$ are normalised to be 1 on the trivial knot.   □

Although this list of properties seems quite comprehensive, our understanding of the new polynomials is much more limited than for the Alexander polynomial. For example, we can construct knots with a given Seifert matrix (Exercise 6.9.14) and hence knots with a given Alexander polynomial. For the Jones polynomial we are unable to do this; we do not even know which polynomials in $\mathbb{Z}[t^{\pm 1}]$ are possible values of the invariant. We have seen that the Alexander polynomial cannot detect knottedness: some non-trivial knots have the same polynomial as the trivial knot. It is not yet known whether similar examples exist for the Jones polynomial, although examples of non-trivial links with the same Jones polynomial as a trivial link have been found. An example is shown in Figure 9.6.

Figure 9.6. This 2-component link has the same Jones polynomial as the trivial 2-component link [175].

9.4   States of graphs and diagrams

Besides giving the axiomatic definition of his bracket polynomial, Kauffman also described a state model, which expresses the invariant as the weighted sum of all possible 'states' of a diagram [188]. One of the benefits of this construction is that it gives bounds on the degree of the polynomial. The main tool, Lemma 9.4.3, is sometimes known as the 'Dual State Lemma'. It can be proved in various ways [19, 188, 200], often using an induction on the number of crossings. The topological argument used here is based on ideas of Vladimir Turaev [208] and uses graphs and surfaces.

Figure 9.7. In a chessboard colouring every vertex has a neighbourhood like this.

Let G be a 4-valent graph embedded in the plane. (Later we shall want to think of G as the underlying projection of a link diagram.) The faces of G can be coloured black and white like a chessboard so that each vertex looks locally like Figure 9.7, and faces that share an edge have different colours (Exercise 3.10.13).

A vertex can be labelled black or white and given a marker to indicate the chosen label as shown on the left of Figure 9.8. A choice of marker for every vertex of G is called a *state*. If G has n vertices then it has 2^n states. A state s has a natural *dual* state, denoted by $\hat{s}$, in which each marker is replaced by the one of the opposite colour. Clearly the dual of $\hat{s}$ is s.

We can remove a labelled vertex from G by splitting it as shown on the right of Figure 9.8. If all the vertices of G are split, the graph becomes a set of disjoint loops in the plane. These loops are called *state circles*. The number of state circles produced by splitting G as indicated by the markers of a state s is denoted by $|s|$.

We shall now construct a closed orientable surface associated to a state s and its dual. Consider $\mathbb{R}^2 \times [-1, 1]$ with G embedded in $\mathbb{R}^2 \times \{0\}$. Outside small neighbourhoods of the vertices, we construct a set of 'walls' of the form $G \times [-1, 1]$. At each vertex we insert a saddle surface (Figure 9.9) positioned in such a way that the boundary curves in $\mathbb{R}^2 \times \{1\}$ correspond to the state circles of s, and the boundary curves in $\mathbb{R}^2 \times \{-1\}$ correspond to the state circles of $\hat{s}$. The state surface is completed by attaching disjoint discs in $\mathbb{R}^3$ to the $|s| + |\hat{s}|$ boundary circles.

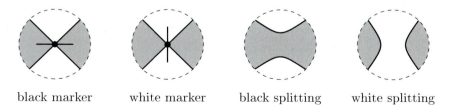

black marker   white marker   black splitting   white splitting

Figure 9.8. The two markers at a vertex and their associated splittings.

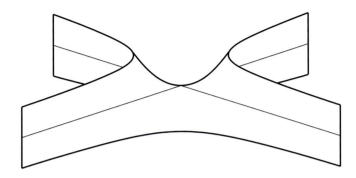

Figure 9.9. A saddle surface can be placed at each vertex.

Lemma 9.4.1. *Let G be a plane connected 4-valent graph, s and $\hat{s}$ two dual states of G, and F the associated state surface. Then, if G has n vertices,*

$$|s| + |\hat{s}| \; = \; 2 + n - 2\,g(F).$$

PROOF. The surface F is connected since G is connected. Because G is 4-valent, it has twice as many edges as vertices, hence $\chi(G) = -n$. The Euler characteristic of the 'walls plus saddles' part of the surface is the same as the Euler characteristic of G. The closed surface has an additional $|s| + |\hat{s}|$ faces so

$$2\,g(F) \; = \; 2 - \chi(F) \; = \; 2 + n - (|s| + |\hat{s}|).$$

Rearranging gives the desired conclusion.   □

A graph is said to be *n-edge-connected* if at least n edges must be deleted to disconnect it. For example, a 2-edge-connected graph does not contain an isthmus. We are interested in the case that G is 3-edge-connected. This implies that the intersection of any pair of adjacent faces is connected (Exercise 9.8.7) and so two faces may meet in a set of vertices, a single edge, or not at all.

Lemma 9.4.2. *Let G be a plane connected 4-valent graph with n vertices, and let s and $\hat{s}$ be two dual states of G. Then*

$$|s| + |\hat{s}| \; \leqslant \; 2 + n.$$

If every state marker in s has the same colour then we get equality. If G is 3-edge-connected and s has markers of both colours, the inequality is strict.

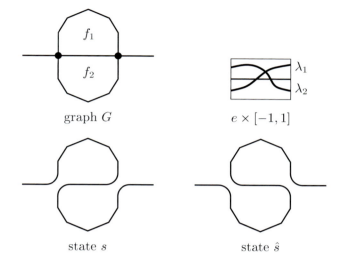

Figure 9.10.

PROOF. The inequality follows from the previous lemma because $g(F) \geqslant 0$. As the only closed orientable surface of genus 0 is a sphere, the result is proved if we show that, under the given conditions, the state surface F is a sphere if and only if all the state markers have the same colour.

Suppose all the markers in s are black. Then the state circles in the lower plane correspond to the black faces of G, and the circles of the dual state in the upper plane correspond to the white faces. Thus F is just the union of all the bounded faces of G plus a disc corresponding to the unbounded face: that is, F is a sphere.

Now suppose that G is 3-edge-connected and there are markers of both colours. There must be an edge e of G that connects a black vertex to a white vertex. Let f_1 and f_2 be the two faces adjacent to e. Because G is even-valent, these faces are distinct; because G is 3-edge-connected, $f_1 \cap f_2 = e$. In the walls of $\partial f_1 \cup \partial f_2$ and saddles at the endpoints of e, choose simple loops λ_1 and λ_2 such that λ_i projects onto ∂f_i. Outside of $e \times [-1, 1]$ these loops are disjoint. However, we can see from Figure 9.10 that λ_1 cannot be embedded entirely in the upper part of the walls $\partial f_1 \times [0, 1]$ — it must drop into $\partial f_1 \times [-1, 0]$ to complete the circuit. This means that λ_1 must intersect e. Similarly λ_2 must also cross e. The two loops are inclined in different directions in $e \times [-1, 1]$ so λ_1 meets λ_2 in an odd number of points (we can assume a single point). This is impossible if F is a sphere.   □

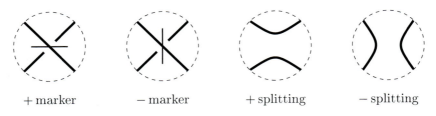

+ marker   − marker   + splitting   − splitting

Figure 9.11. The two markers at a crossing and their associated splittings.

Now we turn from graphs to link diagrams. Let D be a connected diagram of a link L and let G be the underlying projection of D. A marker of G is also a marker of D in a natural way. Because a crossing has more structure than a vertex, it is possible to distinguish two types of marker without the aid of a colouring, as shown in Figure 9.11. Let s_+ and s_- denote the states in which all markers are positive or negative, respectively.

A diagram is *visually prime* if any simple loop in the plane meeting the diagram transversely in exactly two points has crossings on only one side. For example, reducible diagrams are not visually prime; neither is the diagram in Figure 4.3. Note that a diagram of a non-prime link can be visually prime and the opposite (Exercise 9.8.7).

Lemma 9.4.3. *Let D be a connected diagram, and let s and $\hat{s}$ be two dual states of D. Then*

$$|s_+| + |s_-| \leqslant 2 + c(D).$$

If D is alternating then we get equality. If D is non-alternating and visually prime the inequality is strict.

PROOF. Let G be the underlying projection of D. The statement that D is visually prime is equivalent to saying that G is 3-edge-connected (Exercise 9.8.7). Therefore we can apply Lemma 9.4.2 to get $|s_+|+|s_-| \leqslant 2+c(D)$. The markers in s_+ of D have the same colour when interpreted as markers in the coloured graph G if and only if D is alternating. □

Finally, we need to connect these new ideas with the bracket polynomial. For a state s of a diagram D, let $p(s)$ be the number of + markers (or splits) in s and $n(s)$ be the number of − markers. Then $\langle D \rangle$ can be expressed as the following state summation:

$$\langle D \rangle \ = \ \sum_{s} A^{p(s)} (A^{-1})^{n(s)} \delta^{|s(D)|-1}$$

where s ranges over all possible states of D.

9.5 Adequate diagrams

In Chapter 7 we obtained a bound on the degree of the Alexander–Conway polynomial in terms of a geometric invariant (the link genus) and found a class (homogeneous links) on which the bound is always achieved (Corollary 7.6.5). Here we follow a similar path for the bracket polynomial: we obtain bounds on the polynomial and define a class of links for which we can get equality.

Theorem 9.5.1. *For a diagram D*

$$\text{maxdeg}_A \langle D \rangle \leqslant c(D) + 2|s_+(D)| - 2,$$
$$\text{mindeg}_A \langle D \rangle \geqslant -c(D) - 2|s_-(D)| + 2.$$

PROOF. We concentrate on the first inequality, and note that the second can be obtained by a similar analysis, or by considering D^* and using the first.

Suppose we modify a state s_1 by changing a $+$ splitting to a $-$ splitting to produce a new state s_2. Then $p(s_2) = p(s_1) - 1$ and $n(s_2) = n(s_1) + 1$. The number of state circles also changes by 1, and increases or decreases depending on whether the change in state splits one state circle into two, or merges two into one.

For a state s, let $M(s)$ denote the maximum degree of the contribution made to $\langle D \rangle$ by state s. From the state summation formula we get $M(s) = p(s) - n(s) + 2(|s| - 1)$. So

$$
\begin{aligned}
M(s_2) &= p(s_2) - n(s_2) + 2\,|s_2| - 2 \\
&= \big[p(s_1) - 1\big] - \big[n(s_1) + 1\big] + 2\,\big[|s_1| \pm 1\big] - 2 \\
&= M(s_1) - 2 \pm 2 \\
&\leqslant M(s_1).
\end{aligned}
$$

So changing $+$ splittings to $-$ splittings cannot increase $M(s)$.

Therefore no state can contribute a term of degree higher than $M(s_+)$ and

$$\text{maxdeg}_A \langle D \rangle \leqslant M(s_+) = c(D) - 0 + 2\,|s_+(D)| - 2.$$

$\square$

This argument proves the theorem, but we can go further and examine when we get equality. We noted in the proof that the 'plus or minus' ambiguity depends on whether the change from s_1 to s_2 increases or decreases the number of state circles. If the modification increases the number of circles then $M(s_2) = M(s_1)$ and the two terms could reinforce or cancel each other. However, if the number of circles decreases then $M(s_2) < M(s_1)$ and s_2 cannot cancel the highest degree term contributed by s_1. This motivates the following definition [194].

Definition 9.5.2 (adequate diagram). A diagram D is $+$adequate if, at each crossing, the two segments of $s_+(D)$ that replace the crossing are in different state circles. Similarly, D is $-$adequate if the segments of $s_-(D)$ at each crossing are in different circles. If a diagram is neither $+$adequate nor $-$adequate it is called *inadequate*. If a diagram is both $+$adequate and $-$adequate it is called *adequate*.

As usual, a link is *adequate* if it has an adequate diagram, and is called *inadequate* if all its diagrams are inadequate.

Corollary 9.5.3. *For an adequate diagram D*

$$
\begin{aligned}
\mathrm{maxdeg}_A \langle D \rangle &= c(D) + 2|s_+(D)| - 2, \\
\mathrm{mindeg}_A \langle D \rangle &= -c(D) - 2|s_-(D)| + 2.
\end{aligned}
$$

PROOF. Adequacy has been specifically defined to produce the equalities: if D is $+$adequate then s_+ is the unique state that contributes to the highest degree term. This means that the coefficients of the extreme terms are ± 1.
□

One way to prove that a knot is inadequate is to examine the coefficients of the terms of highest and lowest degree. If they are both different from ± 1 then the knot cannot have a $+$adequate diagram or a $-$adequate diagram. This phenomenon does not appear until order 11: an example is shown in Figure 9.12.

Let us now see some examples of adequate diagrams. There is one obvious constraint: for a diagram to be adequate it must be irreducible (the arcs at a nugatory crossing would belong to the same component of either s_+ or s_-).

Theorem 9.5.4. *An irreducible alternating diagram is adequate.*

PROOF. Give the diagram D a chessboard colouring (Exercise 3.10.13). Because D is alternating, the state circles of $s_+(D)$ are loops around the regions of one colour (say black) with the corners smoothed off, and the state circles of $s_-(D)$ are loops around the white regions. Because D is irreducible, no

Figure 9.12. This 11-crossing knot (11n95) is inadequate: the extreme coefficients of its bracket polynomial are ±2.

region meets itself at a crossing, hence none of the state circles visits the same crossing twice and D is adequate. □

The standard diagrams [5] of the non-alternating knots of 8 and 9 crossings are either $+$ adequate or $-$ adequate but not both. Moving up to 10 crossings, we can find non-alternating knots with adequate diagrams. There are three such cases and they are all formed by summing the tangles shown in Figure 9.13 in various ways (see Exercise 9.8.8). These tangles have the following property.

Definition 9.5.5. An alternating diagram of a marked 2-tangle is called *strongly alternating* if both the closures $N(t)$ and $D(t)$ are irreducible.

Theorem 9.5.6. *The non-alternating sum of two strongly alternating tangles is adequate.*

PROOF. Exercise 9.8.10. □

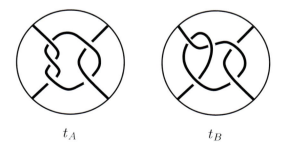

Figure 9.13. The two strongly alternating tangles with five crossings.

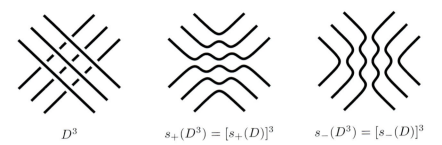

$$D^3 \qquad\qquad s_+(D^3) = [s_+(D)]^3 \qquad s_-(D^3) = [s_-(D)]^3$$

Figure 9.14. Taking the 3-fold parallel of a diagram preserves adequacy.

The diagrams in Figure 7.9 are examples constructed in this way, hence the Kinoshita–Terasaka and Conway knots are adequate.

Another way to construct adequate diagrams is via a 2-dimensional analogue of the satellite construction. The r-fold parallel of a diagram is formed by replacing the single string of each link component by a ribbon of r parallel strings lying flat in the plane of the diagram. The parallel is obtained by making a cable-type satellite of each component of the link in the original diagram. The framing depends on the diagram that is used: adding curls to the original diagram will change the link type of the resulting satellite.

Theorem 9.5.7. *The r-fold parallel of an adequate diagram is adequate.*

PROOF. The proof is essentially contained in Figure 9.14, which shows the case $r = 3$. $\qquad\qquad\square$

This property is extremely useful, as we shall see in the lemma below [204]. First we need some notation.

Given an oriented diagram D, let $c_+(D)$ be the number of positive crossings and $c_-(D)$ be the number of negative crossings. These are related to the crossing number and writhe as follows:

$$
\begin{aligned}
c(D) &= c_+(D) + c_-(D), \\
w(D) &= c_+(D) - c_-(D), \\
c_+(D) &= \tfrac{1}{2}(c(D) + w(D)), \\
c_-(D) &= \tfrac{1}{2}(c(D) - w(D)).
\end{aligned}
$$

Lemma 9.5.8. *Let D be a $+$ adequate diagram of an oriented link. Then $c_-(D)$ is minimal over all diagrams of the link.*

PROOF. We shall prove the lemma for knots and leave its generalisation to links as an exercise (Exercise 9.8.11).

Let D_1 and D_2 be diagrams of an oriented knot K and suppose D_1 is $+$ adequate. We need to show that $c_-(D_1) \leqslant c_-(D_2)$. Define n_1 and n_2 as follows:

$$n_1 = \begin{cases} 0 & \text{if } w(D_1) \geqslant w(D_2), \\ w(D_2) - w(D_1) & \text{if } w(D_1) < w(D_2); \end{cases}$$

$$n_2 = \begin{cases} w(D_1) - w(D_2) & \text{if } w(D_1) > w(D_2), \\ 0 & \text{if } w(D_1) \leqslant w(D_2). \end{cases}$$

Produce a diagram $\widehat{D}_i$ from D_i by inserting n_i positive curls using type 1 Reidemeister moves. Then $w(\widehat{D}_1) = w(\widehat{D}_2)$.

We can consider the r-string parallel $\widehat{D}_i^r$ of $\widehat{D}_i$. Each crossing in $\widehat{D}_i$ gives rise to r^2 crossings in $\widehat{D}_i^r$. Similarly each component of $s_+(\widehat{D}_i)$ becomes r components in $s_+(\widehat{D}_i^r)$. Substituting these into Theorem 9.5.1 we get

$$\mathrm{maxdeg}_A \langle \widehat{D}_2^r \rangle \leqslant \left[c(D_2) + n_2 \right] r^2 + 2 \left[|s_+(D_2)| + n_2 \right] r - 2.$$

The $+$ adequacy of D_1 is preserved both by adding positive curls and by taking parallels. Hence $\widehat{D}_1^r$ is $+$ adequate and we can use the equality of Corollary 9.5.3:

$$\mathrm{maxdeg}_A \langle \widehat{D}_1^r \rangle = \left[c(D_1) + n_1 \right] r^2 + 2 \left[|s_+(D_1)| + n_1 \right] r - 2.$$

By construction, $\widehat{D}_1$ and $\widehat{D}_2$ are diagrams of the same knot K and they have the same writhe. Therefore, the parallels $\widehat{D}_1^r$ and $\widehat{D}_2^r$ are two diagrams of the same link with the same writhe and hence

$$\mathrm{maxdeg}_A \langle \widehat{D}_1^r \rangle = \mathrm{maxdeg}_A \langle \widehat{D}_2^r \rangle.$$

Note that the preceding three relations hold whatever the value of r. By comparing the terms in r^2 we see that

$$c(D_1) + n_1 \leqslant c(D_2) + n_2.$$

Substituting for the n_i and rearranging gives

$$c(D_1) - w(D_1) \leqslant c(D_2) - w(D_2)$$

which implies $c_-(D_1) \leqslant c_-(D_2)$, as required. $\qquad\square$

Corollary 9.5.9. *Let D be a $-$ adequate diagram of an oriented link. Then $c_+(D)$ is minimal over all diagrams of the link.*

PROOF. Switch all the crossings in D and apply the lemma to D^*. □

These two results explain some of the geometry of the Perko pair (Figure 3.12). These are two diagrams of the same knot: 10_{162} is $+$ adequate so $c_-(10_{162}) = 0$ is minimal, and 10^*_{161} is $-$ adequate so $c_+(10^*_{161}) = 9$ is minimal. The two minima cannot be realised simultaneously as the knot does not have a 9-crossing diagram.

The lemma may seem rather obscure and of little interest in itself. However, significant results can be derived from it.

Corollary 9.5.10. *An adequate diagram has minimal crossing number.*

PROOF. If the diagram D is unoriented we choose orientations for its components. An adequate diagram is both $+$ adequate and $-$ adequate and hence both $c_+(D)$ and $c_-(D)$ are minimal. □

This is a very useful result and allows us to find minimal-crossing diagrams 'by inspection'. Without it, proving that a diagram is minimal requires enumerating all diagrams with fewer crossings and proving that they are all different from your given knot, a procedure that can only be done with knots of small crossing number. We can now show that whole families of knot diagrams are minimal, including all irreducible alternating diagrams.

Corollary 9.5.11. *An adequate diagram with at least one crossing represents a non-trivial link.*

In particular, the Kinoshita–Terasaka and Conway knots are non-trivial and have order 11.

Corollary 9.5.12. *Two adequate diagrams of an oriented link have the same crossing number and the same writhe.*

PROOF. Both c_+ and c_- are minimal on adequate diagrams and so are constant. Hence their sum c and difference w are also constant. □

This goes some way to explaining the success of the early tabulators who used the writhe as a link invariant. One consequence of writhe invariance would be that amphicheiral links had even crossing number, but we have seen that this is false in general (see Figure 3.14). It is, however, true for adequate links.

Corollary 9.5.13. *An adequate diagram of an amphicheiral link has zero writhe and even crossing number.*

9.6 Bounds on crossing number

We have seen that adequate diagrams have minimal crossing number, but what can we say about other links? Here we combine results from the previous two sections to derive a lower bound on crossing number in general.

Theorem 9.6.1. *For a diagram D*

$$\operatorname{maxdeg}_A \langle D \rangle - \operatorname{mindeg}_A \langle D \rangle \leqslant 2\,c(D) + 2 \left(|s_+(D)| + |s_-(D)| \right) - 4.$$

If D is adequate then we get equality.

PROOF. This follows immediately from Theorem 9.5.1 and Corollary 9.5.3.
□

This result is not very useful because $|s|$ is not an accessible quantity. Applying the Dual State Lemma (9.4.3) removes this problem.

Theorem 9.6.2. *Let D be a connected diagram. Then*

$$\operatorname{breadth} \widetilde{V}(D) = \operatorname{breadth} \langle D \rangle \leqslant 4\,c(D)$$

If D is alternating and irreducible then we get equality. If D is non-alternating and visually prime the inequality is strict.

PROOF. This follows from Lemma 9.4.3, Theorem 9.5.4 and Theorem 9.6.1. The two polynomials $\widetilde{V}(D)$ and $\langle D \rangle$ differ only by multiplication by a power of A so they have the same breadth.
□

The diagram D of $3_1 \# 3_1^*$ in Figure 1.2 is a non-alternating diagram of an alternating knot but breadth $\widetilde{V}(D) = 24 = 4\,c(D)$. This shows that the requirement that D be visually prime is necessary.

Because $\widetilde{V}$ is a link invariant (unlike $\langle\ \rangle$) we can write

$$\operatorname{breadth} \widetilde{V}(L) \leqslant 4\,c(L).$$

We shall now tidy up a loose end from page 170. Calculation shows that breadth $\widetilde{V}(9_{43}) = 28 = 4 \times 7$. If 9_{43} had an alternating diagram it would have an irreducible alternating diagram with seven crossings. But we have catalogued the knots of order 7 and 9_{43} is different from them all. Therefore 9_{43} is non-alternating. The same argument shows that none of the knots 8_{19}–8_{21} and 9_{42}–9_{49} has an alternating diagram.

9.7   Polynomial link invariants

This section contains a short survey of the polynomial link invariants that have been defined in terms of skein relations. Computers now make it easy to tabulate these.

First we look at the invariants of oriented links. As usual, the links L_+, L_- and L_0 have diagrams D_+, D_- and D_0 which differ as shown in Figure 6.8.

$\Delta_L(t)$: This is the classical Alexander polynomial defined by James Alexander in 1928 [136]. In its original form, it was calculated directly from knot diagrams, a bit like the bracket polynomial. Alexander discovered the following linear relation between different diagrams:

$$\Delta(L_+) - \Delta(L_-) \;=\; (t^{1/2} - t^{-1/2})\,\Delta(L_0).$$

However, in his form, the polynomial was only defined up to powers of t and so adding polynomials of different links together was not a well-defined operation.

The polynomial was reinterpreted and derived in several different ways over the next 50 years. Perhaps the most satisfying of these is from the homology of the branched cyclic covering space of the knot complement. This reveals the underlying geometry and generalises to higher dimensions and to a multivariable version for links.

$\nabla_L(z)$: This is John Conway's potential function, which is a normalised form of the Alexander polynomial and removes the 'powers of t' ambiguity.

Conway rediscovered Alexander's observation and normalised the Alexander polynomial so that it satisfies

$$\nabla(L_+) - \nabla(L_-) \;=\; z\,\nabla(L_0).$$

In the late 1960s, Conway started work on extending the knot tables. His paper [139] contained many new ideas and has had a large influence on the subject. He invented a new method of enumerating link diagrams that can be performed fairly quickly by hand, and enumerated knots with 11 crossings; he introduced a polynomial invariant he called a potential function and the rudiments of skein theory; he introduced the decomposition of links into tangles with their closures, sums, and fraction formula, and a nice relationship between rational tangles and rational numbers that we explored in Chapter 8.

$V_L(t)$: This is the Jones polynomial discovered by Vaughan Jones in the early 1980s [182]. He derived his polynomial invariant via trace functions on von Neumann algebras (see §9.1). Until this point, every method of defining a polynomial invariant had turned out to be just another way of obtaining the Alexander polynomial, so it was something of a surprise to find that Jones' polynomial was a new invariant and not just another derivation of $\Delta(t)$.

Jones worked out that his polynomial satisfied a skein relation:

$$t\,V(L_+) - t^{-1}\,V(L_-) \;=\; (t^{1/2} - t^{-1/2})\,V(L_0).$$

$P_L(v,z)$: After spotting the obvious similarities between the skein relations of the Jones and Alexander polynomials, several people started to work on the general case of a linear skein invariant for oriented links. Five teams — Raymond Lickorish and Ken Millett [193], Jim Hoste [181], Peter Freyd and David Yetter [176], Adrian Ocneanu [201], and Jozef Przytycki and Pawel Traczyk [202] — all reached the same conclusion. The first four submitted announcements of their results to the American Mathematical Society within weeks of each other and the editor asked them to prepare a joint statement [177].

The polynomial is usually called the Homfly polynomial, an acronym formed from the initials of the surnames on the AMS statement. Because of its varied origins, its skein relation appears in several forms which are equivalent under substitution of variables. One of these is

$$v\,P(L_+) - v^{-1}\,P(L_-) \;=\; z\,P(L_0).$$

This is the most general form of a linear skein relation for oriented links.

Just as the Alexander and Jones polynomials give lower bounds on the genus and crossing number respectively, the P polynomial gives lower bounds on the number of strings in a braid, and the canonical genus, as we shall see in Chapter 10.

Now we turn to invariants defined using the unoriented diagrams D_+, D_-, D_0 and D_∞ shown in Figure 9.1.

$\langle D \rangle (A)$: This is the bracket polynomial discovered by Louis Kauffman [188]. Its skein relation is

$$\langle D_+ \rangle \;=\; A \langle D_0 \rangle + A^{-1} \langle D_\infty \rangle.$$

Its axioms do not include invariance under ambient isotopy and it is not a true link invariant: it takes different values on different diagrams of the same link. However, its value is preserved by types 2 and 3 Reidemeister moves. When a type 1 move is applied to the diagram, the bracket polynomial is multiplied by $-A^{\pm 3}$. Kauffman observed that by adding an orientation to the diagram, the writhe can be used to adjust the polynomial to produce an invariant of oriented links. This normalised form is equivalent to the Jones polynomial.

$Q_L(x)$: This is the absolute polynomial discovered by Robert Brandt, Lickorish and Millett [173], and independently by Chi Fai Ho [179, 180]. It is a proper link invariant defined by the skein relation

$$Q(D_+) + Q(D_-) \;=\; x\,[\,Q(D_0) + Q(D_\infty)\,].$$

It is called absolute because it is unaffected by any change of orientation in the link or the ambient space.

$F_L(a, x)$: This is another polynomial discovered by Kauffman [190]. It is a 2-variable generalisation of Q produced by adding an axiom for the case of type 1 Reidemeister moves, and using the writhe to normalise the result. (Because the writhe is involved, the link must be oriented even though the orientation is not used in the skein relation.) The relationship between $Q(x)$ and $F(a, x)$ is analogous to that between $\nabla(z)$ and $P(v, z)$: we regain Q by setting $a = 1$.

Examples have been found to demonstrate the independence of these invariants. For example, there are pairs of knots which have the same P

but different F, and so on. However, there are relationships between some of the polynomials. These are shown below: special choices of variables in P and F produce the polynomials in the bottom row.

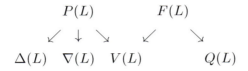

These new invariants are very powerful at distinguishing links, although it is still possible to construct infinite families of links having the same polynomial [185]. None of the polynomials is sensitive to reversal of link orientations; all are unaffected by mutation.

9.8 Exercises

1. Show that the Burau mapping $\mathbb{B}_n \longrightarrow \mathrm{GL}(n, \mathbb{Z}[t])$ is a homomorphism.

2. Show $\widetilde{V}_L(A)$ is a polynomial in A^4 for links with an odd number of components. What about links with even multiplicity?

3. Prove Theorem 9.3.1.

4. Prove Theorem 9.3.2.

5. Show that mutants have the same bracket polynomial.

6. Show that a state surface is orientable.

7. These questions complete the proofs in §9.4.

  (a) Draw a visually prime diagram of a non-prime knot.
  (b) Draw a diagram which is not visually prime but that depicts a prime knot.
  (c) Show that the underlying projection of a diagram is 3-edge-connected if and only if the diagram is visually prime.
  (d) Let G be a 3-edge-connected 4-valent graph. Show that the intersection of any pair of adjacent faces is connected.

8. These questions relate to the tangles shown in Figure 9.13.

  (a) Show that a strongly alternating tangle must have at least five crossings.

(b) Show that, up to symmetry, t_A and t_B are the only strongly alternating tangles with five crossings.

(c) Let t^* denote the reflection of tangle t in the plane of the page, and let $\rho(t)$ denote the tangle t rotated by $90°$ counter-clockwise. Numerators of the following partial sums produce six knots.

Alternating sum		Non-alternating sum	
10_{79}	$t_A \oplus \rho(t_A^*)$	10_{152}	$t_A \oplus \rho(t_A)$
10_{80}	$t_A \oplus \rho(t_B^*)$	10_{153}	$t_A \oplus \rho(t_B)$
10_{81}	$t_B \oplus \rho(t_B^*)$	10_{154}	$t_B \oplus \rho(t_B)$

Show that these are the only possibilities, up to symmetry.

9. Let K be a Lissajous knot with coprime frequencies $B_y > B_x > 1$ and $B_z = 2B_x B_y - B_x - B_y$, and phases $C_x = \pi(2B_x - 1)/(2B_z)$ and $C_y = \pi/(2B_z)$. Show that K has an alternating diagram when viewed along the z-axis [45]. In the example shown below $B_x = 3$ and $B_y = 5$.

(Hint: because the Seifert graph has a single block (Exercise 5.9.15), D is alternating if and only if all the crossings have the same sign.)

Note that this shows there are infinitely many Lissajous knots because irreducible alternating diagrams have minimal crossing number and, by varying the frequencies, we can produce infinitely many with different numbers of crossings.

10. Prove Theorem 9.5.6.

11. Generalise the proof of Lemma 9.5.8 to links. (Define an n for each component of each diagram to adjust the writhes; the contributions from the other crossings are independent of the diagram because linking number is a link invariant.)

10 Closed Braids and Arc Presentations

In this chapter we give space a book-like infrastructure of pages attached to a binding. This enables us to study links in two complementary ways: aligned within the pages or running transversely to them. This provides two new geometric link invariants, lower bounds for which can be obtained from the 2-variable link polynomials $F(a, x)$ and $P(v, z)$.

First, then, let us define the infrastructure. Think of $\mathbb{R}^3$ as $\mathbb{C} \times \mathbb{R}$ with coordinate system (r, θ, z). The z-axis will be the *binding* for the book. Using polar coordinates in the complex plane rather than the standard cartesian ones makes it simpler to describe the *pages*. For a fixed value of θ the set $H_\theta = \{(r, \theta, z) \in \mathbb{R}^3\}$ is the half-plane at angle θ. As θ ranges from 0 to 2π these half-planes fill space — they are the pages of an *open-book decomposition* of $\mathbb{R}^3$.

If we add the point at infinity to form $S^3 = \mathbb{R}^3 \cup \{\infty\}$ then the binding becomes the circle z-axis $\cup \{\infty\}$ and the pages become discs. In both $\mathbb{R}^3$ and S^3, if we remove the binding, we are left with a space homeomorphic to an open solid torus, and the pages are just the meridional discs.

10.1 Braid presentations

In §7.10 we met a special form of n-tangle, called a braid, in which all the strings descend monotonically. Closing the braid produces a link. We can reformulate this idea using the open-book decomposition.

Definition 10.1.1 (closed braid). A link $L \subset \mathbb{R}^3$ is presented as a closed n-string braid if L meets every page H_θ transversely in n points. The binding of the open-book decomposition is called the braid axis. If the link is oriented then all the components must travel around the axis in the same direction.

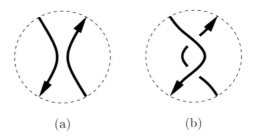

(a)   (b)

Figure 10.1. The reducing move: the two arcs in (a) must belong to
different Seifert circles.

For example, the diagram in Figure 9.5 can be seen as a closed 3-string
braid viewed from a point on the binding.

The fact that every link can be presented as a closed braid was first
proved by James Alexander in 1923 [211]. However, using his method it is
hard to control the complexity of the resulting braid. A modern approach for
producing braids is to take a link diagram and adapt it so that the Seifert
circles are nested one inside the other: when we look down on a closed
braid from a point on the z-axis its Seifert circles will display this pattern.
Shuji Yamada was the first to show that the conversion could always be
done without increasing the number of Seifert circles [244]. Pierre Vogel
found a different method that uses a single type of move — a form of the
standard type 2 Reidemeister move [243]. The following proof that this
move is sufficient is due to Pawel Traczyk [242]: an appropriate choice of
complexity function makes it very concise.

The local move that we need is illustrated in Figure 10.1. The two
oriented arcs in (a) belong to different Seifert circles in the diagram. The
modification which replaces (a) by (b) is called a *reducing move*; it is clear
that it does not affect the link type. The analogous situation in which the
orientations of both arcs are reversed is also allowed.

Theorem 10.1.2. *An oriented link diagram can be transformed into a
closed braid by a finite sequence of reducing moves.*

PROOF. Let $D \subset \mathbb{R}^2$ be a link diagram. We want to think of D as lying in
the 2-sphere $S^2 = \mathbb{R}^2 \cup \{\infty\}$ because then any two Seifert circles bound an
annulus. We also do not need to worry about closures that differ as shown
in Figure 10.2: these diagrams are isotopic on S^2.

Say that two Seifert circles are *compatible* if they have parallel orien-

Figure 10.2. Two diagrams of a closed braid. The links are equivalent; the diagrams are isotopic in S^2 but not in $\mathbb{R}^2$.

tations, meaning that they both travel around the annulus in the same direction. Note that this applies to any two Seifert circles, not just to adjacent ones. For example, all four circles in either diagram in Figure 10.2 are pairwise compatible with each other.

We can use compatibility to define a complexity function that measures how far a diagram is from being a braid. Let $\chi(D)$ be the number of incompatible pairs of Seifert circles in the diagram D taken over all possible pairs. Clearly $\chi(D) \geqslant 0$. Furthermore $\chi(D) = 0$ if and only if all the Seifert circles of D are nested with parallel orientations so that D is a closed braid in S^2.

A reducing move reduces $\chi(D)$ by 1 (Exercise 10.11.3).

Now we need to show that if $\chi(D) > 0$ then a reducing move is possible. The projection underlying D can be thought of as a 4-valent graph embedded in S^2 that divides the sphere into a set of regions or faces. Each edge of the graph is naturally associated with a Seifert circle of D. We say that a face is *reducible* if two of its edges come from incompatible Seifert circles. In this case, a reducing move is possible.

Consider smoothing a crossing in D. The face structure of the graph is changed but the number of reducible faces cannot increase. If we smooth all the crossings we end up with the system of Seifert circles. If some pair of Seifert circles is incompatible, some 'face' in the diagram of 'D with all crossings smoothed' must be reducible, and hence some face of D must also be reducible.

Therefore $\chi(D)$ reducing moves are necessary and sufficient to convert D into a closed braid.   □

Corollary 10.1.3. *Every oriented link can be represented as the closure of a braid $\beta \in \mathbb{B}_n$ for some n.*

This presentation of a link can be used to define another link invariant: the *braid index* of a link L is the smallest n such that L can be presented as a closed n-string braid.

The previous theorem implies that:

Corollary 10.1.4. *The braid index of a link equals the minimum number of Seifert circles in any diagram of the link.*

Therefore, we might equally well call this invariant the Seifert circle index of a link. However, in this form it seems rather contrived, defined just for the sake of it without any real geometric motivation. Braid index is a more natural concept. The fact that these indices are equal means that we can denote the braid index of a link L by $s(L)$. (Recall $b(L)$ is the bridge number.)

Braid index is another of the 'minimise a geometric property' type of link invariants that are extremely difficult to calculate. However, a useful lower bound can be extracted from the Homfly polynomial.

10.2 The Homfly polynomial

The Homfly polynomial is another skein invariant that can be given an axiomatic definition.

Definition 10.2.1. The *Homfly polynomial* of an oriented link L, denoted variously by $P(L)$ or $P_L(v, z)$ depending on the required emphasis, is defined by the three following axioms.

1. Invariance: $P_L(v, z)$ is invariant under ambient isotopy of L.

2. Normalisation: if K is the trivial knot then $P_K(v, z) = 1$.

3. Skein relation: $v^{-1} P(L_+) - v P(L_-) = z P(L_0)$ where L_+, L_- and L_0 have diagrams D_+, D_- and D_0 which differ as shown in Figure 6.8.

It can be computed recursively using a resolving tree, switching and smoothing crossings until the terminal nodes are labelled with trivial links. (This was discussed in Chapter 7 for the Conway polynomial.) In the process, the skein relation is applied in the two following forms:

$$P(L_+) = v^2 P(L_-) + vz P(L_0),$$
$$P(L_-) = v^{-2} P(L_+) - v^{-1}z P(L_0).$$

To complete the calculation we need the following additional relationship, which can be deduced in the same way as Theorem 7.4.2.

4. Delta: $P(L_1 \sqcup L_2) = \delta P(L_1) P(L_2)$ where $\delta = \dfrac{v^{-1} - v}{z}$.

Thus the polynomial of an n-component trivial link is δ^{n-1}.

The invariance of P can be established using either algebraic or combinatorial techniques [177] but there is no short proof and we shall not include one here.

Theorem 10.2.2. *The Homfly polynomial is a well-defined invariant of oriented links that takes values in $\mathbb{Z}[v^{\pm 1}, z^{\pm 1}]$ or $\mathbb{Z}[v^{\pm 1}, z, \delta]$, depending on whether or not we choose to expand the factors of δ.*

Clearly, we can recover the Alexander, Conway, and Jones polynomials from P by substitution:

$$
\begin{aligned}
\Delta_L(x) &= P_L(1, \, x^{-1} - x), \\
\nabla_L(z) &= P_L(1, \, z), \\
V_L(t) &= P_L(t, \, t^{-1/2} - t^{1/2}).
\end{aligned}
$$

We can apply skein theory in the usual way to explain the behaviour of P under the standard link transformations:

Theorem 10.2.3. *The Homfly polynomial has the following properties.*

(a) *If L is a link then $P(-L) = P(L)$.*

(b) *If L is a link then $P(L^*)(v, z) = P(L)(v^{-1}, z)$.*

(c) *If $L_1 \sqcup L_2$ is a split link then $P(L_1 \sqcup L_2) = \delta\, P(L_1)\, P(L_2)$.*

(d) *If a link can be factorised as $L_1 \# L_2$ then $P(L_1 \# L_2) = P(L_1)\, P(L_2)$.*

(e) *If L_1 and L_2 are mutant links then $P(L_1) = P(L_2)$.*

Part (b) shows that the Homfly polynomial of an amphicheiral link must be symmetric in v.

As there is no satellite formula for the Jones polynomial (see the discussion following Theorem 9.3.2), there cannot be one for the Homfly polynomial either.

Theorem 10.2.4. *Let D be a diagram of a μ-component link L. Then*

$$
\begin{aligned}
z\text{-maxdeg}\, P_L(v, z) &\leqslant c(D) - s(D) + 1, \\
z\text{-mindeg}\, P_L(v, z) &\geqslant 1 - \mu(L).
\end{aligned}
$$

PROOF. Let F be the projection surface constructed from D and let G be its Seifert graph. Proceeding in the same way as in the proof of Theorem 7.6.2 we see that z-maxdeg $P_L(v, z) \leqslant \text{rank}(G)$, and, together with Theorem 5.2.1, this suffices to prove the first inequality.

To find the lower bound we need to consider terms involving δ as these are the only source of negative powers of z. The left-most branch of a resolving tree will contribute $\delta^{\mu-1}$ multiplied by a power of v. On other paths from a terminal node to the root of the tree, it is possible that a smoothing operation increases the number of components in the link (so increasing the power of δ) but this is always offset by a power of z that is associated with the smoothing. Hence the net contribution in z^{-1} cannot have a power greater than $\mu - 1$.   □

With a little more work, it can be shown [193] that the lowest degree terms do not cancel and that we get equality for the lower bound: z-mindeg $P_L(v, z) = 1 - \mu$.

Recall that the canonical genus of a link L is the minimum genus of any projection surface spanning L.

Corollary 10.2.5. *The canonical genus of a non-split link L is bounded below by the z-degree of the Homfly polynomial:*

$$2g_c(L) + \mu(L) - 1 \geqslant z\text{-maxdeg} \, P_L(v, z).$$

PROOF. Apply Corollary 5.1.3.   □

Using this result, we can show that genus and canonical genus are distinct link invariants:

Corollary 10.2.6. *The untwisted double of a left-handed trefoil shown in Figure 4.1 does not have a minimal-genus projection surface.*

PROOF. The untwisted double of a trefoil is non-trivial and is spanned by a genus-1 surface, hence it is a genus-1 knot. Its Homfly polynomial can be written in matrix form as

	v^{-10}	v^{-8}	v^{-6}	v^{-4}	v^{-2}	1	v^2
1		1	-5	8	-4		1
z^2	1	1	-10	14	-5	-1	
z^4			-6	7	-1		
z^6			-1	1			

Therefore a projection surface must have genus at least 3. This bound is achievable (Exercise 5.9.13).   □

Notice that when we substitute $v = 1$ into this Homfly polynomial, it collapses to 1, confirming that the Conway polynomial of an untwisted double is 1.

10.3   The Homfly polynomial and braid index

For the Homfly polynomial, we have seen that the degree in z gives a lower bound on the canonical genus. Now we show that the degree in v gives a bound on the braid index. First we need the following lemma [232].

Lemma 10.3.1.   *Let D be an ascending diagram of a (trivial) oriented link with μ components. Then $s(D) + w(D) \geqslant \mu$.*

PROOF. The proof is by induction on the number of crossings in D. The statement is true for diagrams with zero crossings: in this case $w(D) = 0$ and $s(D) = \mu$.

We now consider a diagram with $n > 0$ crossings. Suppose that at some of the crossings both strands are in the same link component. Choose the first such crossing that is encountered when the link is traversed as an ascending diagram (following the ordering of the components and starting each component at its basepoint). We smooth this crossing to produce a diagram D'. Smoothing creates an additional link component but a basepoint can be chosen so that D' is also an ascending diagram (see the proof of Lemma 7.5.1).

Smoothing preserves the number of Seifert circles and changes the writhe by ± 1 depending on the sign of the removed crossing. The number of components of the link has increased by 1. Applying the induction hypothesis to D' gives

$$s(D) + w(D) \pm 1 \; = \; s(D') + w(D') \; \geqslant \; \mu(D') \; = \; \mu(D) + 1$$

from which a simple rearrangement gives the required inequality:

$$s(D) + w(D) \; \geqslant \; \mu(D) + 1 \mp 1 \; \geqslant \; \mu(D).$$

Note that D' is an ascending diagram so the induction is valid.

If there are no self-crossing components then we can suppose that each component is planar and that the components are stacked in parallel planes.

If two components in adjacent planes do not cross each other then we can exchange the levels in which they occur. By doing this if necessary, we can arrange that two components in adjacent planes cross each other. Because the intersection number of these two components is zero, at least one crossing must have positive sign. This time smoothing the crossing reduces the number of components by 1 and decreases the writhe by 1. Hence

$$s(D) + w(D) - 1 \; = \; s(D') + w(D') \; \geqslant \; \mu(D') \; = \; \mu(D) - 1.$$

Again we can arrange that D' is an ascending diagram so the induction is valid. □

Theorem 10.3.2. *Let D be a diagram of a link L. Then*

$$v\text{-maxdeg}\, P_L(v, z) \; \leqslant \; w(D) + s(D) - 1,$$
$$v\text{-mindeg}\, P_L(v, z) \; \geqslant \; w(D) - s(D) + 1.$$

PROOF. First note that we need only prove the first inequality as the second can be derived by applying it to the mirror-image diagram as follows:

$$
\begin{aligned}
v\text{-mindeg}\, P(D)(v, z) \; &= \; -[\, v\text{-maxdeg}\, P(D^*)(v, z)\,] \\
&\geqslant \; -[\, w(D^*) + s(D^*) - 1\,] \\
&= \; w(D) - s(D) + 1.
\end{aligned}
$$

Let $\phi(D) = w(D) + s(D) - 1$. Showing that $v\text{-maxdeg}\, P_L(v, z) \leqslant \phi(D)$ is equivalent to showing that $v\text{-maxdeg}[\, v^{-\phi(D)}\, P_L(v, z)\,] \leqslant 0$.

Suppose that we apply the Homfly skein relation to a positive crossing:

$$P(D_+) \; = \; v^2\, P(D_-) + vz\, P(D_0).$$

Multiplying through by $v^{-\phi(D_+)}$ gives

$$v^{-\phi(D_+)}\, P(D_+) \; = \; v^{-\phi(D_+)}\, v^2\, P(D_-) + v^{-\phi(D_+)}\, vz\, P(D_0)$$

and, noting that $\phi(D_-) = \phi(D_+) - 2$ and $\phi(D_0) = \phi(D_+) - 1$, we can derive

$$v^{-\phi(D_+)}\, P(D_+) \; = \; v^{-\phi(D_-)}\, P(D_-) + v^{-\phi(D_0)}\, z\, P(D_0).$$

So if $P(D_-)$ and $P(D_0)$ satisfy the inequality, $P(D_+)$ will too. A similar argument can be made when we apply the skein relation to a negative crossing.

The proof is completed by using induction on the number of crossings. We convert the given diagram D into an ascending diagram, applying the skein relation as we go to build up the left-most branch of a resolving tree. At the left-most terminal node is an ascending diagram D', and at the other terminal nodes are diagrams with fewer crossings. For the ascending diagram we have v-maxdeg $P_{D'}(v, z) = \mu(D') - 1$. Applying the preceding lemma gives v-maxdeg $P_{D'}(v, z) \leqslant \phi(D')$. The diagrams on the other terminal nodes also satisfy the inequality by the induction hypothesis, so D also satisfies the inequality. $\qquad\square$

The t-breadth of a polynomial is the difference between its highest and lowest degrees in variable t.

Corollary 10.3.3. *The braid index of a link L is bounded below by the v-breadth of the Homfly polynomial:*

$$s(L) \;\geqslant\; \tfrac{1}{2}\, v\text{-breadth}\, P_L(v, z) + 1.$$

This relationship is known as the MFW inequality after its three discoverers: Hugh Morton [232], John Franks and Robert Williams [229]. It holds with equality on all except five knots up to 10 crossings: 9_{42}, 9_{49}, 10_{132}, 10_{150} and 10_{156}.

With a little cunning, we can still use the MFW inequality to determine the braid indices of these exceptional cases. For example: the polynomial bound on the braid index of 9_{42} is 3 but we can only find a 4-string presentation; if it had a 3-string braid presentation then any 2-cable about 9_{42} could be presented on six strings; however, calculating the polynomial of a 2-cable shows its braid index must be 7 or more; hence the braid index of 9_{42} must be at least 4 [149, 198].

10.4 Braid index of rational links

In some cases the MFW inequality is sufficient to establish the braid index on a complete class of links. Here we sketch the argument for rational links.

Let D be the standard diagram of the rational link with continued fraction $C(2a_1, 2a_2, \ldots, 2a_{n-1}, 2a_n)$. Throughout this section it will be convenient to use the sequence $(b_1, b_2, \ldots, b_{n-1}, b_n)$ where $b_i = (-1)^{i+1} 2 a_i$: the signs of these coefficients agree with the signs of the crossings in D. A maximal length subsequence of b_i's that all have the same sign is called a *group*. We shall also use the following notation:

$$\Sigma_+ = \text{sum of the positive } b_i,$$
$$\Sigma_- = \text{sum of the negative } b_i \text{ (so } \Sigma_- < 0),$$
$$\gamma_+ = \text{number of groups of positive } b_i,$$
$$\gamma_- = \text{number of groups of negative } b_i,$$
$$\gamma = \gamma_+ + \gamma_- = \text{number of groups.}$$

Lemma 10.4.1. *Let* $(b_1, b_2, \ldots, b_n)$ *be the sequence just described with corresponding diagram* D. *Then*

$$v\text{-maxdeg } P(D) = \Sigma_+ - n + 2\gamma_+,$$
$$v\text{-mindeg } P(D) = \Sigma_- + n - 2\gamma_-.$$

PROOF (SKETCH). The proof proceeds by induction on n. This is possible because switching and smoothing the left-most crossing in a rational link diagram in which all the twist-boxes contain an even-number of half-twists produces another diagram of the same type.

If we write $P(b_1, \ldots, b_n)$ for the Homfly polynomial of the link formed from this sequence of coefficients, and we assume that $b_n > 0$, then

$$P(b_1, \ldots, b_n) = v^{b_n} P(b_1, \ldots, b_{n-2}) + vz \sum_{i=0}^{b_n/2-1} v^{2i} P(b_1, \ldots, b_{n-1}).$$

The details of the induction argument are left as an exercise. $\square$

Lemma 10.4.2. *Let* L *be the link with diagram* D *and corresponding sequence* $(b_1, b_2, \ldots, b_n)$. *Then*

$$s(L) \geqslant 1 + \gamma - n + \tfrac{1}{2} \sum_{i=1}^{n} |b_i|.$$

PROOF. From the previous lemma we have

$$v\text{-maxdeg } P(D) - v\text{-mindeg } P(D)$$
$$= (\Sigma_+ - n + 2\gamma_+) - (\Sigma_- + n - 2\gamma_-)$$
$$= (\Sigma_+ - \Sigma_-) - 2n + 2(\gamma_+ + \gamma_-)$$
$$= 2\left(\tfrac{1}{2} \sum_{i=1}^{n} |b_i| - n + \gamma\right).$$

Now apply Corollary 10.3.3 to get the lower bound on the braid index. $\square$

We now need to construct a diagram that realises this lower bound. Start with the standard diagram of the rational link shown in Figure 3.9. Take the strand that runs along the top of the diagram and make it pass through the point at infinity (or around the back of the book). This leaves a set of twist-boxes connected by three parallel lines. Labelling the boxes with the b_i's eliminates the negative signs seen in the lower row of Figure 3.9 and simplifies the discussion. The right-hand end of the diagram is shown at the top of Figure 10.3.

Holding the twist-box labelled b_1 fixed, we grab the left side of the diagram and twist it by $180°$. This brings the twist-box labelled b_2 into the upper row; the other twist-boxes also exchange rows. The result is shown in the middle of Figure 10.3. (This is a topological application of the Lagrange formula — see page 204.) It is clear that two of the newly created crossings can be subsumed into the neighbouring twist-boxes, as shown at the bottom of the figure. This procedure has replaced the sequence b_1, b_2, b_3, ... with

$$b_1 - 1, \; -1, \; b_2 - 1, \; b_3, \; \ldots$$

Twisting in the opposite direction produces

$$b_1 + 1, \; 1, \; b_2 + 1, \; b_3, \; \ldots$$

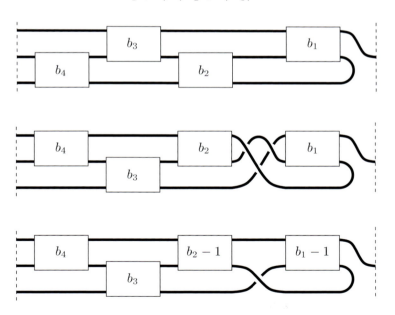

Figure 10.3. Applying a half-twist to a rational link.

By repeating the process and working along the whole diagram, we can obtain a sequence with the following properties:

- the number of terms is odd,
- the even-numbered terms are all ± 1,
- the first and last terms are odd,
- all the other odd-numbered terms are even.

We can choose the sign of the twist between boxes b_i and b_{i+1} so that the number of half-twists in box b_i is reduced. Following this rule, and writing $|b_i|$ for the number of half-twists, we see

$$|b_1| \longmapsto |b_1| - 1;$$

$$|b_i| \longmapsto \begin{cases} |b_i| - 2 & \text{if } \text{sign}(b_i) = \text{sign}(b_{i-1}), \\ |b_i| & \text{if } \text{sign}(b_i) \neq \text{sign}(b_{i-1}); \end{cases}$$

$$|b_n| \longmapsto \begin{cases} |b_n| - 1 & \text{if } \text{sign}(b_n) = \text{sign}(b_{n-1}), \\ |b_n| + 1 & \text{if } \text{sign}(b_n) \neq \text{sign}(b_{n-1}). \end{cases}$$

So, except where there is a change in sign between adjacent b_i's, the number of half-twists can be reduced. The change in sign corresponds to the starting of a new group. To finish the construction, we replace each full-twist in a twist-box with a curl as shown in Figure 10.4. This move reduces two Seifert circles to one each time it is applied. The resulting diagram has the minimal number of Seifert circles.

Figure 10.4. Replacing full-twists reduces the number of Seifert circles.

Proposition 10.4.3. *Let L be the rational link with continued fraction $C(2a_1, 2a_2, \ldots, 2a_{n-1}, 2a_n)$. Let t be the number of times there is a change of sign between adjacent coefficients a_i and a_{i+1}. Its braid index is given by*

$$s(L) = 1 - t + \sum_{i=1}^{n} |a_i|.$$

PROOF. It is straightforward to convert the condition on the b_i coefficients to the equivalent for the a_i. The previous discussion derives the lower bound and shows that it can be achieved. $\quad\square$

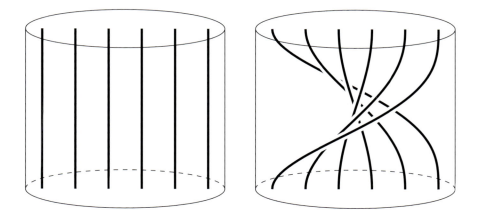

Figure 10.5. Construction of the half-twist Δ in $\mathbb{B}_6$.

10.5 Classification of torus knots

In this section we use the MFW inequality to establish the braid index of a torus knot. When brought together with earlier results, this can be used to find the crossing number of a torus knot, and subsequently complete the classification.

A braid is just a tangle in which the strings descend monotonically. A natural object of this form can be obtained by taking the trivial n-string braid and rotating the bottom of the tangle by $180°$. The resulting braid is called the *half-twist* and is denoted by Δ_n. The construction of Δ_6 is shown in Figure 10.5. The subscript n is often omitted when its value is clear from the context. The braid Δ_n^2 is called the *full-twist*.

The n-string braids form a group $\mathbb{B}_n$ with generators σ_i (see Figure 7.13). There are many ways to write the half-twist in terms of these generators. In particular, we have (Exercise 10.11.2)

$$\begin{aligned} \Delta_n &= (\sigma_1)(\sigma_2\,\sigma_1) \,\cdots\, (\sigma_{n-2} \,\cdots\, \sigma_2\,\sigma_1)\,(\sigma_{n-1} \,\cdots\, \sigma_2\,\sigma_1) \\ \text{and also} \quad \Delta_n &= (\sigma_{n-i+1}\,\sigma_{n-i+2} \,\cdots\, \sigma_{n-1})\,(\sigma_1)\,(\sigma_2\,\sigma_1) \\ &\quad \cdots\, (\sigma_{n-2} \,\cdots\, \sigma_2\,\sigma_1)\,(\sigma_{n-1} \,\cdots\, \sigma_{i+1}\,\sigma_i). \end{aligned}$$

These expressions show that there is a word in $\mathbb{B}_n$ representing Δ_n that finishes with any generator we choose.

A braid word in $\mathbb{B}_n$ is *positive* if only σ_i are used and no inverses σ_i^{-1} appear.

Theorem 10.5.1. *Let $\beta \in \mathbb{B}_n$ be a positive braid and let Δ^2 be the full-twist in $\mathbb{B}_n$. Let L be the link formed as the closure of $\Delta^2 \beta$. Then $s(L) = n$.*

PROOF. We construct a resolving tree for L as follows. Suppose that the first generator in the braid word β is σ_i. Write Δ^2 in a form that terminates with σ_i and let Δ' and β' be such that $\Delta^2 = \Delta' \sigma_i$ and $\beta = \sigma_i \beta'$. Then $\Delta^2 \beta = \Delta' \sigma_i^2 \beta'$.

Applying the Homfly skein relation to σ_i^2 we get

$$
\begin{aligned}
P(\Delta^2 \beta) &= P(\Delta' \sigma_i^2 \beta') \\
&= v^2 P(\Delta' \beta') + vz\, P(\Delta' \sigma_i \beta') \\
&= v^2 P(\Delta' \beta') + vz\, P(\Delta^2 \beta').
\end{aligned}
$$

The braid $\Delta' \beta'$ is a positive braid, and $\Delta^2 \beta'$ is a braid of the form we started with except that it has fewer letters in its braid word. We can apply this principle repeatedly, building the right-hand branch of the resolving tree, until β is empty and we are left with Δ^2 at the right-most node and positive braids at the others.

By Lemma 7.5.1 it is possible to complete the resolving tree by switching and smoothing only positive crossings so that each application of the skein relation is of the form $P(L_+) = v^2 P(L_-) + vz\, P(L_0)$. This means that

$$
P(L) = \sum_i m_i(v, z)\, \delta^{\mu_i - 1}
$$

where the summation is taken over the terminal nodes of the resolving tree, μ_i is the number of components in the trivial link at node i, and $m_i(v, z)$ is the monomial formed as the product of the edge labels on the path from node i to the root of the tree.

Let $M = \max\{\mu_i\}$. The monomials are all positive, so substituting for δ we see that v-breadth $P(L) \geqslant 2M - 2$. One node in the resolving tree is labelled with Δ_n^2. Following the left-most branch from here down to a trivial link, we find a link of n components (we have only switched crossings so the number of components is preserved). Therefore $M \geqslant n$.

Putting the various inequalities together we get

$$
n \;\geqslant\; s(L) \;\geqslant\; \tfrac{1}{2} v\text{-breadth}\, P(L) + 1 \;\geqslant\; \tfrac{1}{2}(2n - 2) + 1 \;=\; n
$$

so $s(L) = n$ as required.   □

An easy corollary gives the braid index of a torus knot:

Proposition 10.5.2. *The braid index of a (p, q) torus knot with $p > q \geqslant 2$ is q.*

PROOF. The (p, q) torus knot has negative crossings so we need to consider its mirror image (which clearly has the same braid index). The braid $(\sigma_1 \sigma_2 \ldots \sigma_{q-1})^p \in \mathbb{B}_q$ is a positive braid presentation of the $(-p, q)$ torus knot. As $q < p$, the braid contains a full-twist and therefore it has minimal braid index. $\qquad\square$

Proposition 10.5.3. *The crossing number of a (p, q) torus knot with $p > q \geqslant 2$ is $p(q - 1)$.*

PROOF. Let D be any diagram of the (p, q) torus knot, K, and let F be the projection surface constructed from D. The previous proposition gives us the braid index of K and Corollary 7.6.7 gives us its genus.

Rearranging the statement of Corollary 5.1.3 gives

$$
\begin{aligned}
c(D) &= 2\,g(F) + s(D) + \mu(D) - 2 \\
&\geqslant 2\,g(K) + s(K) + \mu(K) - 2 \\
&= (p - 1)(q - 1) + q - 1 \\
&= p(q - 1).
\end{aligned}
$$

This bound can be attained: the braid presentation of K described in the previous proposition has this number of crossings. $\qquad\square$

Theorem 10.5.4. *The ambient isotopy classes of torus knots are completely determined by the following relationships:*

- $T(p, q) = T(q, p)$,
- $T(p, q) = T(-p, -q) = -T(p, q)$,
- $T(n, 1)$ *is the trivial knot for all* $n \in \mathbb{Z}$,
- $T(p, q)$ *is non-trivial when* $|p| > |q| \geqslant 2$,
- $T(-p, q)$ *and* $T(p, q)$ *are mirror images and are distinct when* $|p| > |q| \geqslant 2$.

PROOF. First note that the equivalence relationships hold (Exercise 1.12.3). Non-triviality follows because only the trivial knot has braid index 1. To complete the classification we need to show that $T(p, q)$ and $T(r, s)$ are distinct unless (p, q) is equal to one of the following ordered pairs: (r, s), $(-r, -s)$, (s, r), $(-s, -r)$.

For the moment we shall not distinguish between mirror images so we regard $T(r, s)$ and $T(-r, s)$ as equivalent. This means we can assume that $p > q \geqslant 2$ and $r > s \geqslant 2$.

If $T(p, q)$ and $T(r, s)$ are equal then

- their braid indices must be equal so $q = s$,
- their crossing numbers are equal so $p(q - 1) = r(s - 1)$ and hence $p = r$.

It remains to show that $T(p, q)$ and $T(-p, q)$ are distinct. The $(-p, q)$ torus knot has a positive diagram with q Seifert circles and $p(q-1)$ crossings. From Theorem 10.3.2 we obtain

$$v\text{-mindeg}\, P(v, z) \;\geqslant\; w(D) - s(D) + 1 \;=\; (p - 1)(q - 1) \;>\; 0.$$

So the Homfly polynomial cannot be symmetric in v, and hence (using Theorem 10.2.3(b)) torus knots are cheiral.   □

10.6 Arc presentations

Now that we have looked at links running transverse to the pages, we shall switch attention to links running within the pages.

Definition 10.6.1 (arc presentation). An *arc presentation* of a link L is an embedding of L in finitely many pages of the open-book decomposition so that each of these pages meets L in a single simple arc.

An arc presentation of the trefoil is shown in Figure 10.6. The knot is embedded in five half-planes numbered 1 to 5 in sequence around the binding axis; the knot meets the binding in five points, called *vertices*, labelled a to e in sequence along the axis. It is important that the link meets each page in only one arc: if this restriction is removed then three pages always suffice for any link (Exercise 10.11.4). Within a page, the shape of the arc does not matter — see Figure 10.7 for some examples.

The minimum number of pages required to present a given link L in this manner is a link invariant called the *arc index* of L. It is denoted by $\alpha(L)$. We shall show in Theorem 10.7.3 that every link has an arc presentation so this is well-defined.

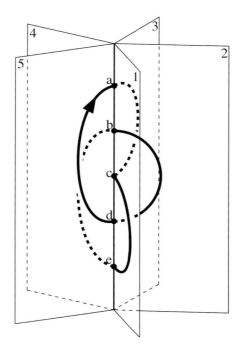

Figure 10.6. A trefoil embedded in five half-planes.

The picture in Figure 10.6 is strongly geometric and three-dimensional, but the same information can be represented in many ways, some of which are shown in Figure 10.8; the same numbers and labels are used in both figures. In the axial form in the top-left the illustration of Figure 10.6 has been simplified by leaving out the sketches of the pages and drawing only the binding and the arcs. In pictures of braids the axis is usually omitted since its location can be inferred; the axis must be included in arc presentations.

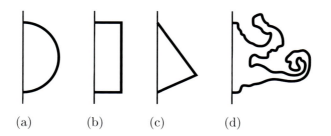

(a)   (b)   (c)   (d)

Figure 10.7. An arc can be any simple path in a page.

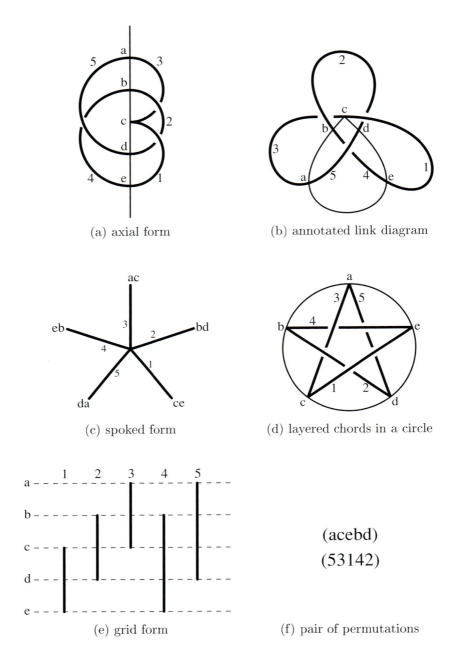

(a) axial form

(b) annotated link diagram

(c) spoked form

(d) layered chords in a circle

(acebd)

(53142)

(e) grid form

(f) pair of permutations

Figure 10.8. Different ways to represent the arc presentation shown in Figure 10.6.

When the construction of Figure 10.6 is viewed from a point on the
binding axis above the knot, we see the axis as a point and each arc as
a spoke emanating from it, as in Figure 10.8(c). In this spoked form, the
labels at the end of each spoke are an essential part of the representation
— the arc presentation cannot be understood without them. In the other
pictorial representations, the letters and numbers are merely cosmetic and
allow us to identify corresponding parts easily.

The embedding can be encoded by a pair of permutations: from a chosen
starting point on the knot, we follow it around noting the order in which
we visit both the half-planes and the vertices. Starting from the arrow, we
get (53142) for the order of the half-planes, and (acebd) for the order of the
points on the axis — Figure 10.8(f). The simplicity of this description has
several advantages:

- It is clear that there are finitely many permutations of n objects so
  there are only finitely many links that can be presented with n arcs.

- It is easy for a computer to enumerate permutations and hence it is
  straightforward to generate a list of all the links that can be presented
  with at most n arcs. This has been done [237] for knots with at most
  nine arcs. The list includes knots with quite large crossing number:
  the (5,4) torus knot with crossing number 15 can be presented with
  nine arcs.

Links that can be presented with few arcs can also be enumerated by
drawing chords in a circle (Exercise 10.11.5). In Figure 10.8(d) the circle
represents the binding and all the arcs are drawn inside. The arcs need not
be straight but they must maintain a strict ordering, lying in layers one
above another.

In Figure 10.8(e) the arcs have been projected onto a tube around the
binding, which has then been cut open and unrolled. To make this statement
more precise, we let $V = D^2 \times A$ be a solid tubular neighbourhood of the
binding axis A, and isotop each arc within its page so that it is composed of
two horizontal lines in $\mathrm{int}(V)$ and a vertical line in ∂V — see Figure 10.7(b).
We now cut open the cylinder ∂V along a line parallel to the arcs (between
arcs 1 and 5) and unroll it: the vertical lines in the figure are the arcs and
the broken horizontal lines mark the heights of their endpoints, $L \cap A$.

Finally, Figure 10.8(b) shows the binding circle drawn onto the usual
diagram of the trefoil. In the next section we shall see how to annotate
diagrams in this manner.

Suppose we have an arc presentation of a link $L \subset \mathbb{C} \times \mathbb{R}$ in our standard (r, θ, z) coordinate system so that the binding A is the z-axis. Let C be a circle in the complex plane, centred at the origin. An arc in page H_θ can be isotoped within the page so that it is formed from two straight line segments that meet at a point on C (see Figure 10.7(c)). When all the arcs of L are arranged in this way, L can be interpreted as an arc presentation in two different ways: the one we started from with binding A and pages H_θ, and also one with binding C and pages that are the cones bounded by C whose apices lie on A. In this case we say that C is the *dual binding*. We shall not make use of the second presentation, except to note that it explains the symmetry between arcs and vertices in what follows.

Definition 10.6.2.

(a) Two arcs are *interleaved* if the endpoints of one alternate with the endpoints of the other around the binding A.

(b) Two vertices are *interleaved* if the arcs meeting at one alternate with the arcs meeting at the other around the dual binding C.

We shall define some operations that can be performed on an arc presentation without affecting the link type. They are illustrated in Figure 10.9. The first two rearrange the order of the arcs or vertices. The other two are analogues of the Markov stabilisation move: they both insert or remove one arc and one vertex.

1. Vertex exchange.
   If two adjacent vertices are not interleaved then they can be interchanged.

2. Arc exchange.
   If two adjacent arcs are not interleaved then they can be interchanged.

3. Vertex merge/divide.
   An arc can be inserted between two arcs with a common endpoint by separating the two arcs along A (adding an extra vertex in the process) and adding a new arc (in its own page) connecting them.

   Conversely, if the two endpoints of an arc are adjacent vertices on A and the neighbouring arcs are disjoint then the arc can be shrunk until it disappears and its endpoints coalesce.

4. Arc merge/divide.

   An arc can be split into two adjacent arcs with the addition of a new vertex.

   Conversely, if the arcs meeting at a vertex are adjacent and do not have the two endpoints in common then they can be merged and their common vertex deleted.

These moves are not independent (one of the stabilisation moves is redundant) but listing all four preserves the symmetry of the situation. Note that move 3 does not specify where the created or deleted arc lies in the page ordering; we could impose the requirement that it is adjacent to one of the other two arcs in the operation and use arc-exchange moves to put it in the desired position. Similarly, in move 4 the created or deleted vertex can lie anywhere along the binding, or we can require that it be adjacent to one of the given ones by using vertex-exchange moves.

Theorem 10.6.3. *Any two arc presentations of a link are related by a finite sequence of the moves shown in Figure 10.9.*

PROOF. Exercise 10.11.10. □

10.7 Arc presentations on diagrams

Let D be the diagram of a link L, and let λ be a simple loop in the plane of the diagram such that $\lambda \cap D$ is a finite set of points away from the crossings in D. The intersections of λ with D need not be transverse — one curve may bounce off the other. We shall investigate the circumstances in which λ can be viewed as the binding of an arc presentation for L.

Take an embedding of L and a regular projection $\pi : \mathbb{R}^3 \longrightarrow \mathbb{R}^2$ that maps L onto the diagram D; the underlying projection $\pi(L)$ is a 4-valent graph G. Let $P = \{p_1, \ldots, p_r\}$ be the set of points where λ meets D. The points $\pi^{-1}(p_i)$ divide L into a set of arcs $A = \{a_1, \ldots, a_r\}$ each of which starts and ends at points in $\pi^{-1}(P)$ and contains no such points in its interior. For each arc a_i in L there is a corresponding part $\pi(a_i)$ of the diagram D which we shall also refer to as an arc.

To see whether a given annotated diagram $D \cup \lambda$ is an arc presentation we can perform the following procedure to colour the arcs of D: the order in which they are coloured determines their sequence around the binding. The plane containing $D \cup \lambda$ is separated by λ into two regions which we

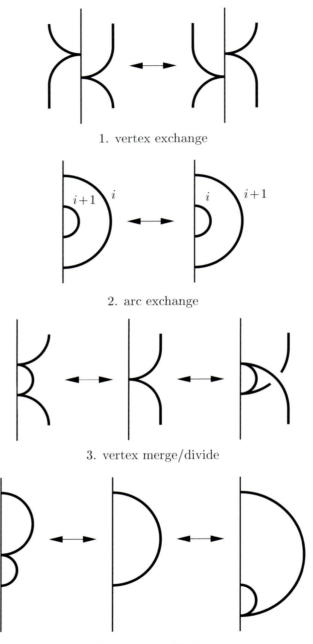

1. vertex exchange

2. arc exchange

3. vertex merge/divide

4. arc merge/divide

Figure 10.9. Equivalence moves on arc presentations [223].

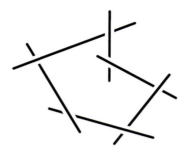

Figure 10.10.

shall refer to as inside and outside. We shall colour the arcs in each region separately, starting with the inside.

The colouring process proceeds under the following rules: an arc can be coloured if

(a) any arc crossing over it is already coloured, and

(b) any arc crossing under it is uncoloured.

The first arc to be coloured is labelled 1. Successive arcs are numbered in the order they are coloured. Repeatedly apply this procedure until it cannot be applied any more. To label the arcs outside λ, we perform the same procedure, labelling the first arc to be coloured by r, continuing by assigning labels in decreasing order.

The colouring process stops when all the arcs have been coloured, or when the conditions cannot be satisfied. An arc cannot be coloured if either

(c) some arc crossing over it is uncoloured, or

(d) some arc crossing under it is already coloured.

Because of the way that the colouring is performed, (d) is impossible. Therefore, if the colouring is unfinished but cannot continue every uncoloured arc must fail condition (c). Consequently, each uncoloured arc must pass under an uncoloured arc. This implies that some of the arcs in D overlay one another in sequence like those shown in Figure 10.10.

We can classify the faces of G as follows.

- A face F is m-type if $\lambda \cap F = \emptyset$: we say λ misses F.

Figure 10.11. A presentation of the (7,3) torus knot with 10 arcs. The construction can be generalised in the obvious way to show that the (p, q) torus knot can be presented with $|p| + |q|$ arcs. This is known to be minimal [228] but the proof is technical and uses surgery on 4-manifolds.

- A face F is t-type if $\lambda \cap \mathrm{int}(F) \neq \varnothing$: we say λ passes through F.

- A face F is b-type if $\lambda \cap \mathrm{int}(F) = \varnothing$ and $\lambda \cap \partial F \neq \varnothing$: we say λ bounces off F.

The set of t-type faces is connected, and every b-type face is adjacent to at least one t-type face. If λ does not miss F we say λ visits F.

Theorem 10.7.1. *If there are no m-type faces then λ is a binding for L.*

PROOF. After the previous discussion, the proof is almost immediate. If λ is not a binding then the colouring procedure will fail to colour all the arcs. The pattern of uncoloured arcs shown in Figure 10.10 divides the plane into two regions and there must be at least one face of G on each side. Because λ lies entirely in one region, there must be at least one m-type face. $\square$

This requirement is not necessary: the annotated diagrams of the torus knot in Figure 10.11 and the pretzel knot in Figure 10.12 do have m-type faces but the colouring procedure shows that both are arc presentations. However, the converse is true for alternating diagrams:

Theorem 10.7.2. *If D is alternating and λ is a binding then there are no m-type faces.*

PROOF. In an alternating diagram, the edges of an m-type face would form a set of arcs like those in Figure 10.10. These arcs cannot be ordered coherently. $\square$

We now show that arc index is a well-defined link invariant.

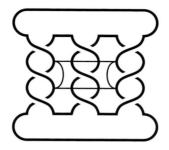

Figure 10.12. A presentation of the $(3, 3, -3)$ pretzel knot with eight arcs.

Theorem 10.7.3. *Every link has an arc presentation. If L is a non-split link then $\alpha(L) \leqslant 2\,c(L) + 2$.*

PROOF. Let D be a diagram of a non-split link L and let G be the 4-valent graph obtained by replacing its crossings with vertices. Because L is non-split, G is connected. Let T be a spanning tree of G and let λ be the boundary of a neighbourhood of T (see Figure 10.13). Then all the faces of G are t-type faces and, by Theorem 10.7.1, λ is a binding.

The loop λ meets every edge of $G - T$ twice, once near each end. Therefore, the number of arcs used is

$$|\lambda \cap D| \;=\; 2\Big(2\,c(D) - (c(D) - 1)\Big) \;=\; 2\,c(D) + 2.$$

If we choose D to be a minimal-crossing diagram then we get a bound on the arc index.

An arc presentation of a split link $L_1 \sqcup L_2$ can be produced by combining presentations for L_1 and L_2 (Exercise 10.11.11(c)). $\qquad\square$

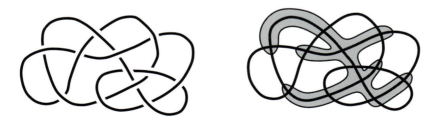

Figure 10.13. A binding can be constructed by drawing round a spanning tree of the underlying graph.

Figure 10.14. For this minimal-crossing diagram, it is not possible to draw a binding circle that visits each face exactly once [216].

This bound is, in fact, rather poor. In the next section we shall see that it is too large by $c(L)$ arcs, almost halving the bound. In many cases we can show that $\alpha(L) \leqslant c(L) + 2$ with an annotated diagram.

Theorem 10.7.4. *If λ visits every face exactly once then the number of arcs is $c(D) + 2$.*

PROOF. To say that λ visits every face once means that $|\lambda \cap F_i| = 1$ for all faces F_i of G: there are no m-type faces, λ bounces once off each b-type face, and λ enters each t-type face once. Therefore $|\lambda \cap D|$ equals the number of faces of G. Since G is 4-valent, we can deduce from Euler's formula that the number of faces equals $c(D) + 2$. $\qquad\square$

Unfortunately, it is not always possible to annotate a diagram so that each face is visited only once. A counterexample is shown in Figure 10.14.

10.8 Arc presentations from diagrams

In this section we describe an algorithm devised by Yongju Bae and Chan-Young Park [212] for converting a link diagram into an arc presentation in spoked form. The algorithm works for all connected diagrams and gives an upper bound in terms of the crossing number. First we have to prove an existence theorem (Lemma 10.8.2) to ensure the bound can be achieved.

A graph is said to be *n-connected* if at least n vertices must be deleted to disconnect it. A 2-connected graph has no cut vertices.

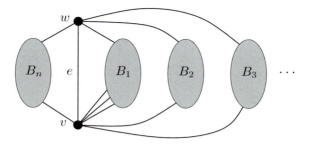

Figure 10.15.

Lemma 10.8.1. *Let G be a 2-connected graph embedded in a plane with no 2-gons. Let G_e be the graph obtained by collapsing the edge e. Given a vertex v there is an edge e incident to v such that G_e is still 2-connected.*

PROOF. Let $e = [v, w]$ be an edge incident to v and suppose that G_e is not 2-connected. Then v is a cut vertex in G_e, and G must have the form shown in Figure 10.15. The subgraphs B_i are blocks in G_e. Each B_i must be connected by at least one edge to both v and w. In block B_1 these edges must be distinct otherwise G would contain a 2-gon bounded by e and another edge forming B_1. (Note that it is possible for block B_2 to contain a single edge connecting v to w.) Choose the edge e_1 in B_1 that is incident to v and nearest to e when turning clockwise around v. Now $G_1 = B_1 \cup e$ is a 2-connected graph with no 2-gons and a chosen vertex v. If G_1 remains 2-connected when e_1 is collapsed then so does G, and e_1 is the desired edge. If not then we have reduced the scale of the problem and can repeat the procedure using G_1 and e_1.

This process cannot be continued indefinitely. The discarded block B_n must contain at least one vertex (as G has no 2-gons) so each time we choose a new subgraph, it has fewer vertices. The only 2-connected graph with no 2-gons and three vertices is a triangle and any of its edges can be used as e. □

Lemma 10.8.2. *Let G be a 2-connected plane graph with more than two vertices. Let G_e° be the graph obtained by collapsing the edge e and removing any innermost loops from the resulting graph. Let v be a vertex (of any valence) and assume that all the other vertices are 4-valent. Then there is an edge e incident to v such that G_e° is still 2-connected.*

PROOF. If a face of G is a 2-gon then delete one of its edges; repeat this process until all the 2-gons have been eliminated. This preserves 2-connectedness so we can apply the previous lemma to select the edge e. Replace the deleted edges to regain G.

The vertex at one end of e is 4-valent so there are four possible cases to consider.

1. If e is not replaced by a multiple edge then we have finished.

2. If e is replaced by a double edge then we can nominate either one as e. When e is collapsed the other edge will become an innermost loop and will be deleted, leaving a 2-connected graph.

3. If e is replaced by a triple edge then we must nominate the middle edge as e. When it is collapsed both the others will become innermost loops.

4. If e were to be replaced by a quadruple edge then the graph would have only two vertices and the four edges, but this is excluded in the statement of the lemma.

$\square$

Theorem 10.8.3.   *If L is a non-split link then $\alpha(L) \leqslant c(L) + 2$.*

PROOF. Take a minimal-crossing diagram D of L. We know that the statement is true when L is the trivial knot so we can assume that $c(D) > 0$.

Choose a crossing c_0 in D and erect a straight line so that it is orthogonal to the plane of the diagram and passes through both the over-crossing and the under-crossing strands of the link at c_0. This line will become the binding axis of the arc presentation. Where the under-crossing strand meets the axis, label both endpoints '1', and where the over-crossing strand meets the axis, label both endpoints '2'.

We shall think of the algorithm from two viewpoints: both in 3-dimensional terms (picking up parts of the link and attaching them to the axis) and in 2-dimensional terms (converting the diagram into an arc presentation in spoked form). At intermediate stages of the process the diagram will be a hybrid object looking like a conventional link diagram with crossing information together with a special vertex that can have high valence and some spokes attached. Each strand of the diagram that ends at the special vertex will be labelled with an integer indicating the level at which it is

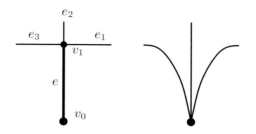

Figure 10.16.

attached to the binding axis. Each spoke will be labelled with a pair of integers indicating the levels of the ends of the arc it represents.

It will also be convenient to associate a graph G with this hybrid representation: G is formed by replacing each crossing in D with a 4-valent vertex and deleting any spokes. The location of the binding axis is marked by a special vertex, denoted by v_0. Because the number of odd-valent vertices in a graph is even, v_0 must be even-valent.

To create the arc presentation we pick up strands of the diagram and attach them to the axis as follows. Select an edge e of G incident to v_0 and let v_1 be the vertex at the other end of e. The four edges meeting at v_1 can be labelled e, e_1, e_2 and e_3 proceeding in counter-clockwise order around the vertex (see the left of Figure 10.16). Let c_1 be the crossing in D that corresponds to v_1. If e is part of the under-crossing strand then take the strand e_1e_3 and attach it to the axis above all the current points and label it with the smallest integer greater than those used so far (see Figure 10.17). Similarly, if e is part of the over-crossing strand then attach the strand e_1e_3 to the axis below all the current points and label it with the largest integer smaller than those already used. The effect of this operation on G is to collapse the edge e forming a new graph G_e (see right of Figure 10.16).

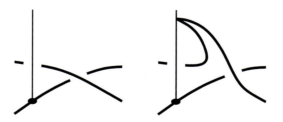

Figure 10.17. An over-crossing strand at a crossing is picked up and attached to the binding axis.

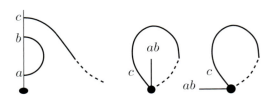

Figure 10.18. Spokes will pass freely under the top-most strand attached to the binding axis.

To create new spokes in the hybrid presentation we need to identify parts of the diagram that can be placed in a vertical plane without affecting the link type. The required parts are loops in G based at v_0 which are innermost in the sense that they contain no other parts of G, although they may contain spokes.

Collapsing e may create 0, 1 or 2 innermost loops. If a loop created in this way contains a spoke, the spoke can be moved outside the loop without affecting the link type (see Figure 10.18). An innermost loop with empty interior can be replaced by a spoke: the spoke is labelled by the pair of integers at the ends of the loop.

Having outlined the procedure we wish to follow, we must now show that it can always be applied without leading to an impasse. It is clear that, while G has vertices other than v_0 (so that D has crossings), there will always be an edge e and an over- or under-crossing strand to attach to the axis. If collapsing e to give G_e produces innermost loops then these can be exchanged for spokes.

The one thing that must be avoided during the construction is the creation of loops based at v_0 that are not innermost loops. Such loops will be labelled at both ends and so should correspond to a spoke, but it may not be possible to force the strand of the link to lie in a plane and become an arc while also preserving the link type and the planarity of the arcs already formed. Additional arcs must be introduced into the presentation to eliminate such loops and we are trying to use a minimal number of arcs. Non-innermost loops can arise as soon as e is collapsed or later if v_0 becomes a cut vertex. We can prevent these cases by using Lemma 10.8.2 to select the edges.

The eight configurations for the edges e, e_1, e_2 and e_3 and their connections to v_0 are shown in Figure 10.19.

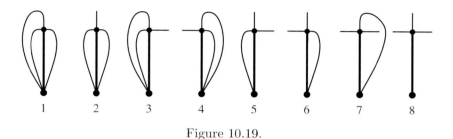

Figure 10.19.

In case 8 no innermost loops are created so we get one new label but no extra spokes. In cases 5 and 6 one label and one innermost loop are created, which leads to one extra spoke. In case 2, a new label and two loops/spokes are created.

Cases 3 and 4 are not possible: Lemma 10.8.2 always selects the middle edge of a triple edge. Case 7 would also produce a non-innermost loop but it, too, will never be chosen because v_0 would then become a cut vertex.

This leaves case 1. Here there are only two vertices so Lemma 10.8.2 does not apply; we choose e to be one of the edges not in the boundary of the unbounded region, as in Figure 10.19. The situation is initially like case 2: two innermost loops are produced and they are exchanged for two spokes. What remains is a hybrid diagram with no crossings so we cannot choose a vertex v_1 and an edge e. We just choose a midpoint on the remaining loop and attach it to the axis above all the current points and label it accordingly. This produces two innermost loops, both of which can be converted into spokes as before. At this point the hybrid contains only spokes and we have an arc presentation of L.

How many arcs have we used in this arc presentation? Notice that collapsing an edge does not alter the number of faces in G. Also one face is exchanged for one spoke when we delete an innermost loop. Therefore the number of faces plus the number of spokes is preserved at each step of the construction until we reach the penultimate stage when there are no crossings left. At the start G has $c(D)+1$ (bounded) faces and no spokes, and at the penultimate stage there is only one face so there must be $c(D)$ spokes. The final reduction converts the last loop into two spokes, producing an arc presentation for L with $c(D) + 2$ arcs.   □

It is known that this construction gives a minimal arc presentation when D is an alternating diagram [224]. Furthermore, it is possible to extend the argument to show that if D is non-alternating then $\alpha(L) < c(D) + 2$ (that is the inequality is strict) [212, 214].

10.9   Braid presentations of satellites

In this section we return to the study of closed braids, and in particular to the construction of braid presentations for satellite links. This discussion has been deferred because, as we shall see, in some cases arc presentations play an essential role in the construction.

Our objective is to find a closed braid presentation for a satellite S and relate its structure and complexity to properties of the pattern $P \subset W$, companion knot C, and framing f. The ease with which this can be done depends crucially on a simple property of the pattern: whether or not the wrapping number and the winding number are equal.

Let L be a link in a (possibly knotted) solid torus V. Suppose that there is some meridional disc of V that meets L in m points, all of which have the same sign of intersection. Thus the wrapping number and the winding number of L are both equal to m. We say that L is a *one-way* link in V. If the wrapping number and winding number are different then every meridional disc of V will have intersections with L of opposite sign: in this case we say that L is *bidirectional* in V. Both patterns and satellites can be one-way or bidirectional.

On the left of Figure 10.20 is a diagram of a pattern $P \subset W$ formed from a closed n-string braid β arranged inside a torus: the shaded regions contain parallel strings without crossings. Suppose that m of the braid strings pass around the tube of the torus so that some meridional disc of W meets P in m points, all of which have the same sign of intersection. Thus P is a one-way pattern. The spot in the figure, next to the braid box, marks the location of the braid axis. If $m < n$ then the axis meets the torus in two points; if $m = n$ then there are no strings in the inner shaded region and the axis can be moved to pass through the hole of the torus and not intersect it. When this pattern is used to form a satellite, we can control the framing by using a faithful homeomorphism and adjusting the power of the full-twist, Δ_m^2, that appears in the pattern.

Theorem 10.9.1.   *Let K_f be the satellite constructed from a one-way pattern $P \subset W$ with framing f and companion C and suppose that the pattern has the form shown on the left of Figure 10.20. Then the braid index of the satellite is bounded above:*

$$s(K_f) \ \leqslant \ m\,s(C) + (n - m).$$

Note that this is independent of the framing.

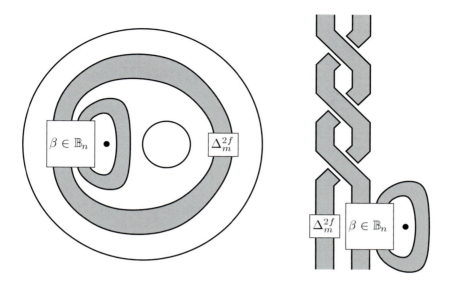

Figure 10.20. Construction of a closed braid presentation of a one-way satellite (when the wrapping number and winding number are equal).

PROOF. Take a closed braid presentation of C on the minimum number of strings, and let V be a tubular neighbourhood of C. Let $h : W \longrightarrow V$ be a faithful homeomorphism. Then $h(P) = K_f$: the power of Δ_m^2 in the pattern can be varied to produce the desired framing. An example satellite is shown on the right of Figure 10.20: in this case the companion knot is a trefoil (the top and bottom of the picture must be identified to close the braid).

Let A be the braid axis. The m strings of P that run round W become m strings that run round the satellite. Because C is presented as a braid, these strings are also braided about A. The remaining $n - m$ strings are sitting inside the torus V and do not yet go round the axis. To convert $h(P)$ into a closed braid we must isotop the torus so that ∂V meets the axis in two points, and then move the $n - m$ strings through the axis. The number of strings used in this construction is $m\,s(C) + (n - m)$. $\qquad\square$

This construction is straightforward and is known to produce a minimal braid [107, 220, 237]. When the wrapping number and the winding number of the pattern are different, there are strings of P running around W in both directions and another approach is required. We shall illustrate the difficulties with a case study of the doubles of a trefoil.

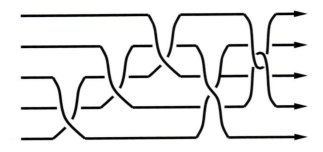

Figure 10.21. Braid presentation for a double of the trefoil.

Figure 10.21 shows a braid presentation of a double of a left-handed trefoil. It is easier to identify this satellite when the braid is closed, as shown in Figure 10.22, and drawn as though it is the boundary of a surface formed from five horizontal discs connected by vertical bands. The surface is then seen to be an annulus, knotted in the form of a trefoil, with a second twisted annulus attached — the genus-1 spanning surface of a double. In this form, it is also clear that the crossings in the four bands on the left control the framing of the satellite and that by switching the signs of these crossings, we can adjust the framing. What determines the framing is the relative number of positive and negative crossings in these bands; the distribution across the bands is irrelevant. Each type of crossing can be used on 0, 1, 2, 3, or 4 bands, giving five possibilities: in fact a braid of this form exists for the five framings in $\{-6, \ldots, -2\}$.

To extend the framing range we need to add extra strings to the braid. The way that the extra string is added determines whether the range is extended at the upper or lower end. Figure 10.23 shows two more braid

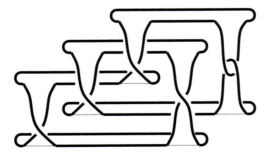

Figure 10.22. A closed braid spanned by discs and bands.

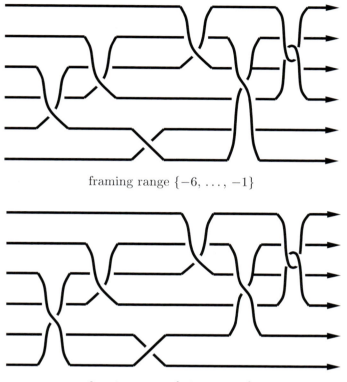

framing range $\{-6, \ldots, -1\}$

framing range $\{-7, \ldots, -2\}$

Figure 10.23. Two ways to augment the braid presentation of Figure 10.21.

presentations of the same double as shown in Figure 10.21 (the braid closures are ambient isotopic and the framing is the same in all three cases). However, in one case the additional crossing is positive, and in the other it is negative. Switching the additional crossing will thus produce a different effect on the framing in the two cases. (Recall that to calculate the framing we use two strings running around the torus in the same direction whereas the strings at the switched crossing run in opposite directions. Therefore, switching a positive crossing to a negative crossing in the diagram actually increases the framing.)

The way that the number of braid strings required by this construction depends on the framing is shown in Figure 10.24. We can prove that the construction gives a minimal braid:

Proposition 10.9.2. *Let K_f be the satellite constructed from the White-head pattern (see left of Figure 4.1) with framing f and companion 3_1, the left-handed trefoil. Then the braid index of the double is given by*

$$
s(K_f) = \begin{cases}
-f - 1 & \text{if } f < -6, \\
5 & \text{if } -6 \leqslant f \leqslant -2, \\
f + 7 & \text{if } -2 < f.
\end{cases}
$$

PROOF. The knot K_f is shown in Exercise 5.9.13 where $n = -(f+3)$.

We have seen a construction which gives an upper bound of the required form. We can obtain a lower bound using the MFW inequality (Corollary 10.3.3).

Consider the Homfly polynomial evaluated at $z = 1$; this gives a Laurent polynomial in v. It is possible that some terms may cancel, allowing the breadth to drop, but we still have

$$
2(s(L) - 1) \;\geqslant\; v\text{-breadth } P_L(v, z) \;\geqslant\; v\text{-breadth } P_L(v, 1).
$$

The table in Figure 10.25 shows the general behaviour of $P(v, 1)$ for the knot K_f: column 1 contains the value of $n = -(f+3)$; column 3 lists the coefficients of the polynomial $P(v, 1)$; column 2 indicates the power of v corresponding to the 'central' coefficient of the polynomial $(-22, -20, \text{ or } -21)$; and column 4 contains the lower bound on the braid index derived from the breadth of $P(v, 1)$. $\square$

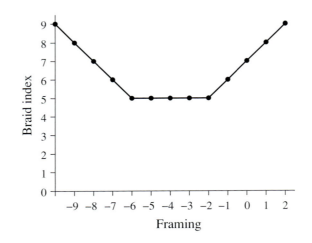

Figure 10.24. Braid index versus framing for doubles of the trefoil.

n	Power				Coefficients of $P(v,1)$						Bound
$-m$	v^{-2m}			1	2	-22	30	-10	$\overbrace{-1\ \cdots\ -1}^{m-2}$	1	$m+4$
$\vdots$	$\vdots$				$\vdots$						$\vdots$
-3	v^{-6}			1	2	-22	30	-10	-1	1	7
-2	v^{-4}			1	2	-22	30	-10	-1		6
-1	v^{-2}			1	2	-22	30	-8			5
0	1			1	2	-22	32	-9			5
1	v^2			1	2	-20	31	-9			5
2	v^4			1	4	-21	31	-9			5
3	v^6			3	3	-21	31	-9			5
4	v^8		2	2	3	-21	31	-9			6
5	v^{10}	2	1	2	3	-21	31	-9			7
$\vdots$	$\vdots$				$\vdots$						$\vdots$
m	v^{2m}	2	$\overbrace{1\ \cdots\ 1}^{m-4}$	2	3	-21	31	-9			$m+2$

Figure 10.25.

This example shows that the braid index of a bidirectional satellite can depend on the framing. In fact, all satellites with wrapping number 2 and winding number 0 exhibit the same kind of behaviour — a similar form of lower bound can be derived from the MFW inequality [237, 239].

What about bidirectional satellites with wrapping number greater than two? Notice that the braid presentation in Figure 10.21 has a strong resemblance to the grid form of arc presentation shown in Figure 10.8(e). We can use this observation to construct braid presentations of all bidirectional satellites.

Suppose that we want to construct a braid presentation for a satellite S which has the knot C as a companion. Take an arc presentation of C with n arcs and binding axis A. Let $W = D^2 \times S^1$ be a solid torus and let $h : W \longrightarrow \mathbb{R}^3$ be a homeomorphism such that $V = h(W)$ is a solid tubular neighbourhood of C with $h(\{0\} \times S^1) = C$. Let $\{p_1, \ldots, p_n\} = C \cap A$ be the set of points where C meets the binding axis. We can assume that h is such that each component of $A \cap V$ lies inside a meridional disc of the form $h(D^2 \times h^{-1}(p_i))$, and that these discs are planar. These discs slice V into a series of solid cylinders, each being the tubular neighbourhood of an arc. We want to construct a closed braid that runs through these cylinders (and hence has ∂V as a companion torus) and has A for its braid axis.

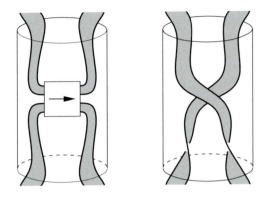

Figure 10.26. A braid inside a cylinder (left) and a completion tube.

On the left of Figure 10.26 is a braid inside a cylinder; as before, the shaded regions represent parallel strings without crossings. Although the strings run through the cylinder in both directions, they progress strictly from left to right across the page. The numbers of strings entering and leaving the braid box must be equal, but the numbers of strings at the top and bottom of the cylinder may be different. By placing a braid in each of the n sections of V we can build a link that is braided about the binding axis and has C as a companion. Drawing the arc presentation in grid form makes this easy to see: in Figure 10.27 the companion knot is (again) a left-handed trefoil, and the braid boxes are numbered in their sequence along the torus.

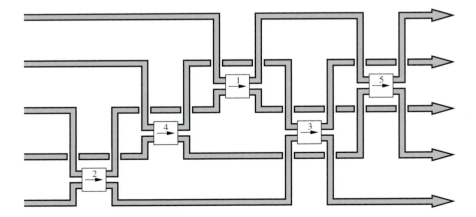

Figure 10.27. A braid presentation of a bidirectional satellite with trefoil companion.

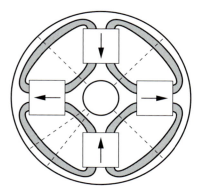

Figure 10.28. A bidirectional pattern with four braid boxes.

What does the pattern for this satellite look like? The discs $D^2 \times h^{-1}(p_i)$ divide the torus W into sections. Placing braid boxes in these sections produces a pattern like that shown in Figure 10.28: the broken lines indicate the sections. Notice that there are five sections in the torus, but we can only use an even number of braid boxes. Where has the fifth braid gone?

The subtle difference between Figures 10.27 and 10.28 is that the former has a natural framing imposed by the arc presentation (see Exercise 10.11.12) and the latter is unframed. The fifth section in the satellite has the form shown on the right of Figure 10.26: the strings are given a half-twist as they pass through the cylinder. This can be expressed as a braid: if the two ribbons have p and q strings, respectively, then the braid contains p parallel strings crossing over q parallel strings followed by two adjacent half-twist braids, Δ_p and Δ_q. By exchanging this braid for its inverse, we change the framing of the satellite by ± 1. Sections of this form are called *completion tubes*. Whenever there are fewer braid boxes in the pattern than there are arcs in the presentation, the deficit is made up with these completion tubes: each one provides an opportunity to alter the framing of the satellite.

So, to construct a braid for a given satellite S with companion C and bidirectional pattern $P \subset W$, we can use the following procedure.

1. Arrange the pattern as a sequence of r braid boxes (like the example shown in Figure 10.28).

2. Take an arc presentation of C with n arcs where $n \geqslant r$.

3. Add $n - r$ extra sections to the pattern to form completion tubes.

4. Map each section of W onto a solid tubular neighbourhood of an arc.

5. If the required framing is outside the range available by switching the crossings in the completion tubes, increase n.

Although this recipe does not provide a simple formula for the braid index of a bidirectional satellite in terms of its components, it suggests that the braid index should depend on the following factors: the number and size of the braid boxes, the arc index of the companion, the natural framing induced by the arc presentation, and the framing of the satellite.

10.10   Applications of the open-book infrastructure

Although we have used the open-book infrastructure only as a way of defining particular presentations of links, it was introduced into knot theory in a significant way as a tool for studying how closed braids interact with surfaces of various kinds. In a series of papers [218, 219, 220], Joan Birman and William Menasco studied the following cases:

- a splitting sphere,
- a factorising sphere,
- a companion torus,
- a spanning surface, particularly a disc spanning the trivial knot.

When the braid axis is used as the binding, the pages in the open-book decomposition intersect the given surface in a pattern known as a *foliation*. The pattern of intersections allows a tiling of the surface to be defined, and, by examining the local arrangements of the tiles, we can make deductions about its position relative to the braid axis. In certain circumstances, the braid can be rearranged so that the embedding of the surface can be simplified; eventually, the surface can be placed in a canonical position. The important thing is that the complexity of the braid is controlled throughout the process — the number of strings is not allowed to increase at any time.

In §9.1 we noted that any equivalent braids are connected by a finite sequence of Markov moves. However, one of these moves increases the number of braid strings and, unfortunately, in some cases use of this move cannot be avoided: there are braid presentations of the trivial knot on four strings that cannot be simplified without it [231]. Rather than use Markov moves, Birman and Menasco found that they could achieve the necessary simplifications by applying what is known as an *exchange move* (see Figure 10.29):

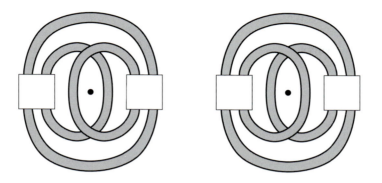

Figure 10.29. The exchange move on a braid.

this does not change number of braid strings but can change the isotopy class of the link in the complement of the braid axis. By applying finite sequences of these moves, they showed that:

- Any closed braid of a split link can be arranged so that the splitting sphere meets the axis twice.

- Any closed braid of a product link can be arranged so that the factorising sphere meets the axis twice. This means that the factorisation becomes visible with braids of the two factors arising naturally. The solution to a long standing conjecture about the additivity of braid index is an easy consequence:

$$s(L_1 \,\#\, L_2) \;=\; s(L_1) + s(L_2) - 1.$$

- Any closed braid of a one-way satellite can be arranged so that the companion torus meets the axis at most twice. This shows that the construction described in Theorem 10.9.1 gives the best result possible.

- Any closed braid of a bidirectional satellite can be arranged so that the companion torus is a tubular neighbourhood of an arc presentation of the companion knot.[*]

Over the years, arc presentations have appeared in the literature several times in different guises:

---

[*]In [220] some cases were overlooked; the analysis was completed by Birman's student Ka Yi Ng [236].

- In one of the earliest papers on knot theory [221], Hermann Brunn asked whether it was possible to have a projection of any link with a single singular point of high multiplicity. His solution was to construct something similar to an arc presentation and then look along the axis.

- Lee Neuwirth [235] used them to give presentations of the fundamental group of the link complement in which the relators do not contain inverse powers of the generators. He called them 'asterisk presentations'.

- A link diagram can be drawn so that all the lines are horizontal or vertical, and the over-crossing strands are always vertical. Such diagrams are similar to the grid form of an arc presentation. Herbert Lyon [230] called them square-bridge presentations and used them in 3-manifold constructions. Lee Rudolph [241] called them fences (the vertical lines being the posts and the horizontal ones the wires) in his study of quasipositivity. They are also related to the Legendrian knots and links in contact structures.

- Arc presentations have also been used to obtain upper bounds on the rope length of knots (the minimum length of rope with unit diameter needed to construct the knot) [222].

In each case, arc presentations were used only as a tool in a proof: they were not seen as being of interest in their own right. This changed when they were implicated in the satellite construction.

The basic properties of the new invariant, arc index, were developed by Peter Cromwell and Ian Nutt. Cromwell introduced moves analogous to the Reidemeister and Markov moves, and adapted Birman and Menasco's technique to show how arc index behaves on links that can be split or factorised [223]:

$$\alpha(L_1 \sqcup L_2) = \alpha(L_1) + \alpha(L_2),$$
$$\alpha(L_1 \# L_2) = \alpha(L_1) + \alpha(L_2) - 2.$$

Nutt made a computer enumeration of arc presentations of knots with up to nine arcs. He also proved a counterpart of the MFW inequality, relating the arc index to the breadth of the Kauffman polynomial:

$$\alpha(L) \geqslant a\text{-breadth } F_L(a, x) + 2.$$

Nutt managed to prove this for knots only with the coefficients of F reduced modulo 2 [238]; the full version was proved by Hugh Morton and Elisabetta Beltrami using the layered chords form of the arc presentation [233].

Together, Cromwell and Nutt developed the colouring algorithm described in §10.7 for constructing a binding circle on a diagram [225]. Empirical evidence led them to conjecture that $\alpha(L) \leqslant c(L) + 2$ with equality if and only if L is alternating. The original algorithm applied to a large class of diagrams (including the rational links and all the knots up to nine crossings except 9_{40}) and gave support to the conjecture. Beltrami and Cromwell extended the algorithm but eventually proved that a 2-dimensional approach could not work in all cases [215, 216]. Yongju Bae and Chan-Young Park introduced the 3-dimensional spoked diagram construction described in §10.8, giving the expected upper bound on arc index: $\alpha(L) \leqslant c(L) + 2$ [212]. When combined with the polynomial lower bound and the fact [207] that $c(L) = a$-breadth $F_L(a, x)$ for alternating links, this showed that $\alpha(L) = c(L) + 2$ for alternating links. They suggested a modification to their method to reduce the number of arcs used in the construction when starting from a non-alternating diagram; Beltrami extended their work, confirming that the inequality is strict for non-alternating links [214].

The open-book infrastructure can also be used to solve the knot triviality problem. Recall that the solutions discussed in §3.9 (Haken's algorithm based on 3-manifolds, and the Hass–Lagarias bound on the number of Reidemeister moves required) are not useful in practice. Birman and Menasco applied the foliated surface technique to the problem. By studying the disc spanning a closed braid of the trivial knot, they showed that it can always be simplified to the trivial 1-string braid using only exchange moves and Markov's stabilisation move that reduces the number of strings [219]. Although the complexity of the braid (the number of strings) never increases, this approach is still not useful because exchange moves can be performed repeatedly to produce infinitely many different braids. The same approach was applied to arc presentations by Ivan Dynnikov [226].* By studying a disc spanning an arc presentation of the trivial knot, he showed that it can be reduced to a presentation with two arcs using only arc/vertex exchange moves, and arc/vertex merge operations to reduce the number of arcs. In this case, it is clear that an algorithm which tries all possible moves will terminate in finite time because there are only finitely many possibilities at each step: if no move is possible, the knot must be non-trivial. The last statement provides another way to show that knots are non-trivial 'by inspection'. For example, it is not possible to apply arc/vertex exchange operations or arc/vertex merge operations to the trefoil in Figure 10.6; therefore it must be knotted.

---

*This paper also corrects an oversight in [218], an error inherited by [223], in the proofs of the additivity theorems.

10.11 Exercises

1. Let $K(p,q)$ denote Taizo Kanenobu's family of knots shown below.

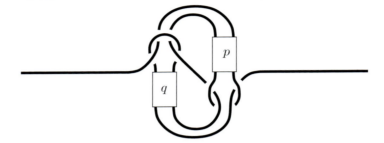

Show

(a) $K(0,0) = 4_1 \# 4_1$,
(b) $K(p,q) = K(q,p)$,
(c) $K(-p,-q) = K(p,q)^*$,
(d) $K(p,q)$ is a ribbon knot,
(e) the Homfly polynomial of $K(p,q)$ depends only on $p+q$,
(f) $K(p,q)$ is amphicheiral if and only if $p = -q$,
(g) $K(p,q)$ has genus 2.

In fact, except for the symmetry in part (b), all the knots in the family are distinct (see [185]). Thus part (e) provides infinitely many knots with the same Homfly polynomial.

2. Show the braid words on page 253 correspond to the half-twist Δ_n.

3. Complete the proof of Theorem 10.1.2: show that a reducing move reduces $\chi(D)$ by 1.

4. Show that any link can be embedded in a book with three pages if the link is allowed to pass through a page more than once.

5. By enumerating the possibilities show that

(a) the trivial knot is the only link with $\alpha = 2$,
(b) no link has $\alpha = 3$,
(c) the trivial 2-component link and the Hopf link are the only links with $\alpha = 4$.

(Hint: use the layered chords format.)

6. Show $p(L) \leqslant 2\alpha(L) - 1$.

7. Using the grid form of the arc presentation, fill in parts of the broken horizontal lines in different ways to show that

(a) $\alpha(L) \geqslant 2\, s(L)$,
(b) $\alpha(L) \geqslant 1 + 2\sqrt{c(L)}$.

8. Annotate the diagrams of the rational links in Figure 3.9 to produce a binding with $c(L) + 2$ arcs.

9. Interpret the moves of Figure 10.9 in each of the styles shown in Figure 10.8.

10. Prove Theorem 10.6.3. (Hint: make a link diagram or a braid from the grid form of arc presentation, then show that the Reidemeister moves or the Markov moves can be simulated using the moves of Figure 10.9 [223, 226].)

11. Show by construction that

(a) $s(L_1 \sqcup L_2) \leqslant s(L_1) + s(L_2)$,
(b) $s(L_1 \# L_2) \leqslant s(L_1) + s(L_2) - 1$,
(c) $\alpha(L_1 \sqcup L_2) \leqslant \alpha(L_1) + \alpha(L_2)$,
(d) $\alpha(L_1 \# L_2) \leqslant \alpha(L_1) + \alpha(L_2) - 2$.

(Equality is known to hold in all cases — see §10.10.)

12. Take an arc presentation in grid format of a knot C. Let S be the bidirectional 2-string cable of C constructed by placing the elementary braid $\sigma_1 \in \mathbb{B}_2$ in a tubular neighbourhood of each arc (see Figure 10.27 and the description in the text). Show that the framing of the satellite equals the writhe of the grid diagram.

13. A *plait* is formed from a braid with an even number of strings by closing it with cups and caps, as shown below. Show that every link can be represented as the plait-closure of a positive braid.

14. Prove the statement in the caption of Figure 10.14.

Appendix A   Knot Diagrams

There is a scheme for determining whether an alternating cheiral knot is left- or right-handed [27, 46]. The letter in brackets indicates the cheirality of the depicted knot: L, D and A denote laevo, dextro and amphicheiral respectively. Orientations are taken from [27]; if the diagram is not oriented the knot is reversible.

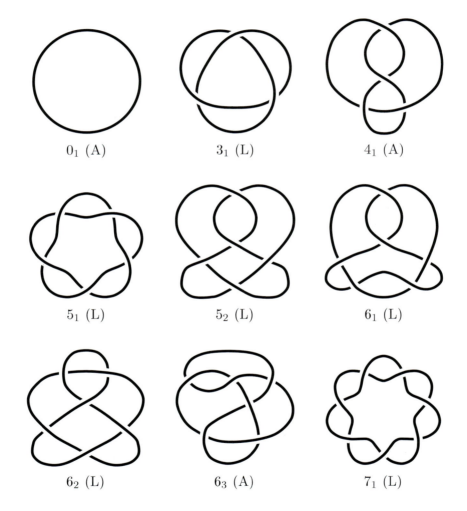

| 0_1 (A) | 3_1 (L) | 4_1 (A) |

| 5_1 (L) | 5_2 (L) | 6_1 (L) |

| 6_2 (L) | 6_3 (A) | 7_1 (L) |

7_2 (L)

7_3 (D)

7_4 (D)

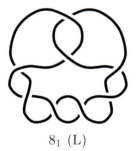
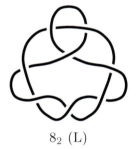

7_5 (L)

7_6 (L)

7_7 (D)

8_1 (L)

8_2 (L)

8_3 (A)

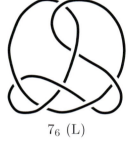
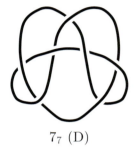
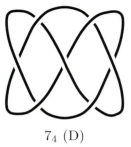
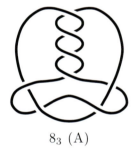

8_4 (D)

8_5 (D)

8_6 (L)

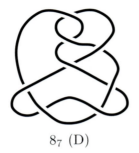

8_7 (D)

8_8 (D)

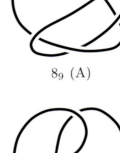

8_9 (A)

8_{10} (D)

8_{11} (L)

8_{12} (A)

8_{13} (D)

8_{14} (L)

8_{15} (L)

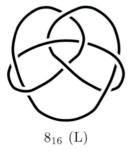

8_{16} (L)

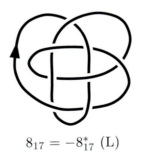

$8_{17} = -8_{17}^{*}$ (L)

8_{18} (A)

8_{19}

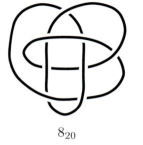

8_{20}

8_{21}

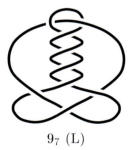

9_1 (L)

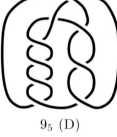

9_2 (L)

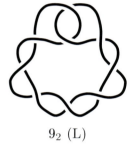

9_3 (D)

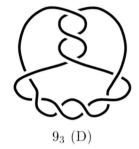

9_4 (L)

9_5 (D)

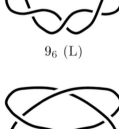

9_6 (L)

9_7 (L)

9_8 (L)

9_9 (L)

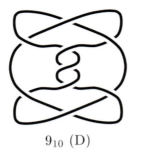

9_{10} (D)

9_{11} (D)

9_{12} (L)

9_{13} (D)

9_{14} (D)

9_{15} (D)

9_{16} (D)

9_{17} (L)

9_{18} (L)

9_{19} (L)

9_{20} (L)

9_{21} (D)

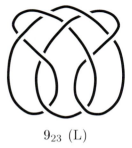

9_{22} (D) 9_{23} (L) 9_{24} (L)

9_{25} (L) 9_{26} (D) 9_{27} (L)

9_{28} (L) 9_{29} (L) 9_{30} (L)

9_{31} (L) 9_{32} (D) 9_{33} (L)

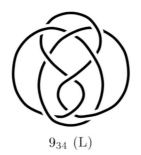

9_{34} (L)

9_{35} (L)

9_{36} (D)

9_{37} (L)

9_{38} (L)

9_{39} (D)

9_{40} (L)

9_{41} (L)

9_{42}

9_{43}

9_{44}

9_{45}

9_{46} 9_{47} 9_{48}

9_{49}

Appendix B   Numerical Invariants

This table lists the numerical invariants discussed in the text.

Polygon indices are taken from [55]. Richard Randell proved the minimality of the polygons with up to eight edges [54], so also did Seiya Negami [94]; Jorge Calvo proved the 9-edge polygons to be minimal [245]; the others are candidates for minimal polygons.

Genus is verified using the degree of the Alexander polynomial — see Theorem 7.2.1.

Determinant and signature are calculated as described in §6.6.

Unknotting numbers can be determined using the signature bound of Corollary 6.8.3 or from the fact that a value of 1 for a non-trivial knot must be minimal. The values marked † have been determined using other methods [249, 251]. When the upper bound found by experiment and the lower bound obtained from theory do not agree both are given. Most of the values are taken from [18]; several cases have been resolved since that table was compiled: 8_{16} and 9_{49} [253], 9_{17} [246], and 8_{10}, 9_{29} and 9_{32} [252].

Bridge numbers are either 1 (the trivial knot), 2 (for rational knots) or 3.

Braid indices are verified using the MFW inequality (see Corollary 10.3.3 and the subsequent discussion).

Arc indices are from [213, 237].

K	$p(K)$	$g(K)$	$\det(K)$	$|\sigma(K)|$	$u(K)$	$b(K)$	$s(K)$	$\alpha(K)$
0_1	3	0	1	0	0	1	1	2
3_1	6	1	3	2	1	2	2	5
4_1	7	1	5	0	1	2	3	6
5_1	8	2	5	4	2	2	2	7
5_2	8	1	7	2	1	2	3	7
6_1	8	1	9	0	1	2	4	8
6_2	8	2	11	2	1	2	3	8
6_3	8	2	13	0	1	2	3	8
7_1	9	3	7	6	3	2	2	9
7_2	9	1	11	2	1	2	4	9
7_3	9	2	13	4	2	2	3	9
7_4	9	1	15	2	$2^\dagger$	2	4	9
7_5	9	2	17	4	2	2	3	9

K	$p(K)$	$g(K)$	$\det(K)$	$|\sigma(K)|$	$u(K)$	$b(K)$	$s(K)$	$\alpha(K)$
7_6	9	2	19	2	1	2	4	9
7_7	9	2	21	0	1	2	4	9
8_1	10	1	13	0	1	2	5	10
8_2	10	3	17	4	2	2	3	10
8_3	10	1	17	0	$2^\dagger$	2	5	10
8_4	10	2	19	2	$2^\dagger$	2	4	10
8_5	10	3	21	4	2	3	3	10
8_6	10	2	23	2	$2^\dagger$	2	4	10
8_7	10	3	23	2	1	2	3	10
8_8	10	2	25	0	$2^\dagger$	2	4	10
8_9	10	3	25	0	1	2	3	10
8_{10}	10	3	27	2	$2^\dagger$	3	3	10
8_{11}	10	2	27	2	1	2	4	10
8_{12}	10	2	29	0	$2^\dagger$	2	5	10
8_{13}	10	2	29	0	1	2	4	10
8_{14}	10	2	31	2	1	2	4	10
8_{15}	10	2	33	4	2	3	4	10
8_{16}	9	3	35	2	$2^\dagger$	3	3	10
8_{17}	9	3	37	0	1	3	3	10
8_{18}	9	3	45	0	$2^\dagger$	3	3	10
8_{19}	8	3	3	6	3	3	3	7
8_{20}	8	2	9	0	1	3	3	8
8_{21}	9	2	15	2	1	3	3	8
9_1	10	4	9	8	4	2	2	11
9_2	11	1	15	2	1	2	5	11
9_3	11	3	19	6	3	2	3	11
9_4	10	2	21	4	2	2	4	11
9_5	10	1	23	2	$2^\dagger$	2	5	11
9_6	11	3	27	6	3	2	3	11
9_7	10	2	29	4	2	2	4	11
9_8	10	2	31	2	$2^\dagger$	2	5	11
9_9	10	3	31	6	3	2	3	11
9_{10}	10	2	33	4	2 or 3	2	4	11
9_{11}	11	3	33	4	2	2	4	11
9_{12}	10	2	35	2	1	2	5	11
9_{13}	10	2	37	4	2 or 3	2	4	11
9_{14}	10	2	37	0	1	2	5	11
9_{15}	11	2	39	2	$2^\dagger$	2	5	11

K	$p(K)$	$g(K)$	$\det(K)$	$|\sigma(K)|$	$u(K)$	$b(K)$	$s(K)$	$\alpha(K)$
9_{16}	10	3	39	6	3	3	3	11
9_{17}	10	3	39	2	$2^{\dagger}$	2	4	11
9_{18}	11	2	41	4	2	2	4	11
9_{19}	10	2	41	0	1	2	5	11
9_{20}	10	3	41	4	2	2	4	11
9_{21}	11	2	43	2	1	2	5	11
9_{22}	10	3	43	2	1	3	4	11
9_{23}	11	2	45	4	2	2	4	11
9_{24}	10	3	45	0	1	3	4	11
9_{25}	11	2	47	2	$2^{\dagger}$	3	5	11
9_{26}	10	3	47	2	1	2	4	11
9_{27}	11	3	49	0	1	2	4	11
9_{28}	10	3	51	2	1	3	4	11
9_{29}	9	3	51	2	$2^{\dagger}$	3	4	11
9_{30}	10	3	53	0	1	3	4	11
9_{31}	10	3	55	2	$2^{\dagger}$	2	4	11
9_{32}	10	3	59	2	$2^{\dagger}$	3	4	11
9_{33}	10	3	61	0	1	3	4	11
9_{34}	9	3	69	0	1	3	4	11
9_{35}	10	1	27	2	2 or 3	3	5	11
9_{36}	11	3	37	4	2	3	4	11
9_{37}	10	2	45	0	$2^{\dagger}$	3	5	11
9_{38}	10	2	57	4	2 or 3	3	4	11
9_{39}	10	2	55	2	1	3	5	11
9_{40}	9	3	75	2	$2^{\dagger}$	3	4	11
9_{41}	9	2	49	0	$2^{\dagger}$	3	5	11
9_{42}	9	2	7	2	1	3	4	8
9_{43}	10	3	13	4	2	3	4	9
9_{44}	9	2	17	0	1	3	4	9
9_{45}	10	2	23	2	1	3	4	9
9_{46}	9	1	9	0	$2^{\dagger}$	3	4	8
9_{47}	9	3	27	2	$2^{\dagger}$	3	4	9
9_{48}	10	2	27	2	$2^{\dagger}$	3	4	9
9_{49}	9	2	25	4	$3^{\dagger}$	3	4	9

Appendix C   Properties

This table lists the properties discussed in the text. All the non-trivial knots listed are simple (and hence prime).

The column marked c indicates the cheirality of the knot: the symbol $\pm$ means that $K = \pm K^*$; the symbol $-$ means that $K = -K^*$ but $K \neq K^*$; the symbol $\times$ means K is cheiral.

The next five columns indicate whether the knot is reversible (invertible), alternating, positive, homogeneous, or ribbon. Within the range of the table, all the adequate knots are alternating so column 'a' also indicates adequacy. Figure 4.8 shows ribbon presentations of 6_1 and 8_{20}; diagrams for the others can be found in Appendix F5 of [18].

If a knot can be represented as a torus knot, pretzel knot, or a rational knot then an example is shown. The fraction corresponding to the rational knot is also listed.

K	c	i	a	p	h	r	p/q	Families		
0_1	$\pm$	✓	✓	✓	✓	✓	$1/0$	$T(1,0)$	$P(1,1,-1)$	$C(0,0)$
3_1	$\times$	✓	✓	✓	✓	$\times$	$3/1$	$T(3,2)$	$P(1,1,1)$	$C(3)$
4_1	$\pm$	✓	✓	$\times$	✓	$\times$	$5/2$	$P(2,1,1)$	$C(2,2)$	
5_1	$\times$	✓	✓	✓	✓	$\times$	$5/1$	$T(5,2)$	$C(5)$	
5_2	$\times$	✓	✓	✓	✓	$\times$	$7/3$	$P(3,1,1)$	$C(2,3)$	
6_1	$\times$	✓	✓	$\times$	✓	✓	$9/4$	$P(4,1,1)$	$C(2,4)$	
6_2	$\times$	✓	✓	$\times$	✓	$\times$	$11/4$	$C(2,1,3)$		
6_3	$\pm$	✓	✓	$\times$	✓	$\times$	$13/5$	$C(2,1,1,2)$		
7_1	$\times$	✓	✓	✓	✓	$\times$	$7/1$	$T(7,2)$	$C(7)$	
7_2	$\times$	✓	✓	✓	✓	$\times$	$11/5$	$P(5,1,1)$	$C(2,5)$	
7_3	$\times$	✓	✓	✓	✓	$\times$	$13/4$	$P(-5,2,1)$	$C(3,4)$	
7_4	$\times$	✓	✓	✓	✓	$\times$	$15/4$	$P(3,3,1)$	$C(3,1,3)$	
7_5	$\times$	✓	✓	✓	✓	$\times$	$17/7$	$C(2,2,3)$		
7_6	$\times$	✓	✓	$\times$	✓	$\times$	$19/7$	$C(2,1,2,2)$		
7_7	$\times$	✓	✓	$\times$	✓	$\times$	$21/8$	$C(2,1,1,1,2)$		
8_1	$\times$	✓	✓	$\times$	✓	$\times$	$13/6$	$P(6,1,1)$	$C(2,6)$	
8_2	$\times$	✓	✓	$\times$	✓	$\times$	$17/6$	$C(2,1,5)$		
8_3	$\pm$	✓	✓	$\times$	✓	$\times$	$17/4$	$P(-5,3,1)$	$C(4,4)$	
8_4	$\times$	✓	✓	$\times$	✓	$\times$	$19/5$	$C(3,1,4)$		
8_5	$\times$	✓	✓	$\times$	✓	$\times$		$P(3,3,2)$		
8_6	$\times$	✓	✓	$\times$	✓	$\times$	$23/10$	$C(2,3,3)$		

K	c	i	a	p	h	r	p/q	Families
8_7	×	✓	✓	×	✓	×	23/9	$C(2,1,1,4)$
8_8	×	✓	✓	×	✓	✓	25/9	$C(2,1,3,2)$
8_9	±	✓	✓	×	✓	✓	25/7	$C(3,1,1,3)$
8_{10}	×	✓	✓	×	✓	×		
8_{11}	×	✓	✓	×	✓	×	27/10	$C(2,1,2,3)$
8_{12}	±	✓	✓	×	✓	×	29/12	$C(2,2,2,2)$
8_{13}	×	✓	✓	×	✓	×	29/11	$C(2,1,1,1,3)$
8_{14}	×	✓	✓	×	✓	×	31/12	$C(2,1,1,2,2)$
8_{15}	×	✓	✓	✓	✓	×		
8_{16}	×	✓	✓	×	✓	×		
8_{17}	−	×	✓	×	✓	×		
8_{18}	±	✓	✓	×	✓	×		
8_{19}	×	✓	×	✓	✓	×		$T(4,3)$ $P(3,3,-2)$
8_{20}	×	✓	×	×	×	✓		$P(3,-3,2)$
8_{21}	×	✓	×	×	×	×		
9_1	×	✓	✓	✓	✓	×	9/1	$T(9,2)$ $C(9)$
9_2	×	✓	✓	✓	✓	×	15/7	$P(7,1,1)$ $C(2,7)$
9_3	×	✓	✓	✓	✓	×	19/6	$P(-7,2,1)$ $C(3,6)$
9_4	×	✓	✓	✓	✓	×	21/5	$P(-6,3,1)$ $C(4,5)$
9_5	×	✓	✓	✓	✓	×	23/6	$P(5,3,1)$ $C(3,1,5)$
9_6	×	✓	✓	✓	✓	×	27/11	$C(2,2,5)$
9_7	×	✓	✓	✓	✓	×	29/13	$C(2,4,3)$
9_8	×	✓	✓	×	✓	×	31/11	$C(2,1,4,2)$
9_9	×	✓	✓	✓	✓	×	31/9	$C(3,2,4)$
9_{10}	×	✓	✓	✓	✓	×	33/10	$C(3,3,3)$
9_{11}	×	✓	✓	×	✓	×	33/14	$C(2,2,1,4)$
9_{12}	×	✓	✓	×	✓	×	35/13	$C(2,1,2,4)$
9_{13}	×	✓	✓	✓	✓	×	37/10	$C(3,1,2,3)$
9_{14}	×	✓	✓	×	✓	×	37/14	$C(2,1,1,1,4)$
9_{15}	×	✓	✓	×	✓	×	39/16	$C(2,2,3,2)$
9_{16}	×	✓	✓	✓	✓	×		
9_{17}	×	✓	✓	×	✓	×	39/14	$C(2,1,3,1,2)$
9_{18}	×	✓	✓	✓	✓	×	41/17	$C(2,2,2,3)$
9_{19}	×	✓	✓	×	✓	×	41/16	$C(2,1,1,3,2)$
9_{20}	×	✓	✓	×	✓	×	41/15	$C(2,1,2,1,3)$
9_{21}	×	✓	✓	×	✓	×	43/18	$C(2,2,1,1,3)$
9_{22}	×	✓	✓	×	✓	×		
9_{23}	×	✓	✓	✓	✓	×	45/19	$C(2,2,1,2,2)$

K	c	i	a	p	h	r	p/q	Families
9_{24}	×	✓	✓	×	✓	×		
9_{25}	×	✓	✓	×	✓	×		
9_{26}	×	✓	✓	×	✓	×	47/18	$C(2,1,1,1,1,3)$
9_{27}	×	✓	✓	×	✓	✓	49/19	$C(2,1,1,2,1,2)$
9_{28}	×	✓	✓	×	✓	×		
9_{29}	×	✓	✓	×	✓	×		
9_{30}	×	✓	✓	×	✓	×		
9_{31}	×	✓	✓	×	✓	×	55/21	$C(2,1,1,1,1,1,2)$
9_{32}	×	×	✓	×	✓	×		
9_{33}	×	×	✓	×	✓	×		
9_{34}	×	✓	✓	×	✓	×		
9_{35}	×	✓	✓	✓	✓	×		$P(3,3,3)$
9_{36}	×	✓	✓	×	✓	×		
9_{37}	×	✓	✓	×	✓	×		
9_{38}	×	✓	✓	✓	✓	×		
9_{39}	×	✓	✓	×	✓	×		
9_{40}	×	✓	✓	×	✓	×		
9_{41}	×	✓	✓	×	✓	✓		
9_{42}	×	✓	×	×	×	×		
9_{43}	×	✓	×	×	✓	×		
9_{44}	×	✓	×	×	×	×		
9_{45}	×	✓	×	×	×	×		
9_{46}	×	✓	×	×	×	✓		$P(3,3,-3)$
9_{47}	×	✓	×	×	✓	×		
9_{48}	×	✓	×	×	✓	×		
9_{49}	×	✓	×	✓	✓	×		

Appendix D Polynomials

The polynomials are calculated using the Morton–Short program [150, 255]. For knots up to order 8 the Jones polynomials are for the diagrams shown in Appendix A; for order 9 the polynomials with fewest negative powers are printed (so some may be mirror images). The Alexander and Conway polynomials are not sensitive to cheirality.

Alexander polynomial

3_1 $x^{-2} - 1 + x^2$

4_1 $-x^{-2} + 3 - x^2$

5_1 $x^{-4} - x^{-2} + 1 - x^2 + x^4$

5_2 $2x^{-2} - 3 + 2x^2$

6_1 $-2x^{-2} + 5 - 2x^2$

6_2 $-x^{-4} + 3x^{-2} - 3 + 3x^2 - x^4$

6_3 $x^{-4} - 3x^{-2} + 5 - 3x^2 + x^4$

7_1 $x^{-6} - x^{-4} + x^{-2} - 1 + x^2 - x^4 + x^6$

7_2 $3x^{-2} - 5 + 3x^2$

7_3 $2x^{-4} - 3x^{-2} + 3 - 3x^2 + 2x^4$

7_4 $4x^{-2} - 7 + 4x^2$

7_5 $2x^{-4} - 4x^{-2} + 5 - 4x^2 + 2x^4$

7_6 $-x^{-4} + 5x^{-2} - 7 + 5x^2 - x^4$

7_7 $x^{-4} - 5x^{-2} + 9 - 5x^2 + x^4$

8_1 $-3x^{-2} + 7 - 3x^2$

8_2 $-x^{-6} + 3x^{-4} - 3x^{-2} + 3 - 3x^2 + 3x^4 - x^6$

8_3 $-4x^{-2} + 9 - 4x^2$

8_4 $-2x^{-4} + 5x^{-2} - 5 + 5x^2 - 2x^4$

8_5 $-x^{-6} + 3x^{-4} - 4x^{-2} + 5 - 4x^2 + 3x^4 - x^6$

8_6 $-2x^{-4} + 6x^{-2} - 7 + 6x^2 - 2x^4$

8_7 $x^{-6} - 3x^{-4} + 5x^{-2} - 5 + 5x^2 - 3x^4 + x^6$

8_8 $2x^{-4} - 6x^{-2} + 9 - 6x^2 + 2x^4$

8_9 $-x^{-6} + 3x^{-4} - 5x^{-2} + 7 - 5x^2 + 3x^4 - x^6$

8_{10} $x^{-6} - 3x^{-4} + 6x^{-2} - 7 + 6x^2 - 3x^4 + x^6$

8_{11} $-2x^{-4} + 7x^{-2} - 9 + 7x^2 - 2x^4$

8_{12} $x^{-4} - 7x^{-2} + 13 - 7x^2 + x^4$

8_{13} $2x^{-4} - 7x^{-2} + 11 - 7x^2 + 2x^4$

8_{14} $-2x^{-4} + 8x^{-2} - 11 + 8x^2 - 2x^4$

8_{15} $3x^{-4} - 8x^{-2} + 11 - 8x^2 + 3x^4$

8_{16} $x^{-6} - 4x^{-4} + 8x^{-2} - 9 + 8x^2 - 4x^4 + x^6$

8_{17} $-x^{-6} + 4x^{-4} - 8x^{-2} + 11 - 8x^2 + 4x^4 - x^6$

8_{18} $-x^{-6} + 5x^{-4} - 10x^{-2} + 13 - 10x^2 + 5x^4 - x^6$

8_{19} $x^{-6} - x^{-4} + 1 - x^4 + x^6$

8_{20} $x^{-4} - 2x^{-2} + 3 - 2x^2 + x^4$

8_{21} $-x^{-4} + 4x^{-2} - 5 + 4x^2 - x^4$

9_1 $x^{-8} - x^{-6} + x^{-4} - x^{-2} + 1 - x^2 + x^4 - x^6 + x^8$

9_2 $4x^{-2} - 7 + 4x^2$

9_3 $2x^{-6} - 3x^{-4} + 3x^{-2} - 3 + 3x^2 - 3x^4 + 2x^6$

9_4 $3x^{-4} - 5x^{-2} + 5 - 5x^2 + 3x^4$

9_5 $6x^{-2} - 11 + 6x^2$

9_6 $2x^{-6} - 4x^{-4} + 5x^{-2} - 5 + 5x^2 - 4x^4 + 2x^6$

9_7 $3x^{-4} - 7x^{-2} + 9 - 7x^2 + 3x^4$

9_8 $-2x^{-4} + 8x^{-2} - 11 + 8x^2 - 2x^4$

9_9 $2x^{-6} - 4x^{-4} + 6x^{-2} - 7 + 6x^2 - 4x^4 + 2x^6$

9_{10} $4x^{-4} - 8x^{-2} + 9 - 8x^2 + 4x^4$

9_{11} $-x^{-6} + 5x^{-4} - 7x^{-2} + 7 - 7x^2 + 5x^4 - x^6$

9_{12} $-2x^{-4} + 9x^{-2} - 13 + 9x^2 - 2x^4$

9_{13} $4x^{-4} - 9x^{-2} + 11 - 9x^2 + 4x^4$

9_{14} $2x^{-4} - 9x^{-2} + 15 - 9x^2 + 2x^4$

9_{15} $-2x^{-4} + 10x^{-2} - 15 + 10x^2 - 2x^4$

9_{16} $2x^{-6} - 5x^{-4} + 8x^{-2} - 9 + 8x^2 - 5x^4 + 2x^6$

9_{17} $x^{-6} - 5x^{-4} + 9x^{-2} - 9 + 9x^2 - 5x^4 + x^6$

9_{18} $4x^{-4} - 10x^{-2} + 13 - 10x^2 + 4x^4$

9_{19} $2x^{-4} - 10x^{-2} + 17 - 10x^2 + 2x^4$

9_{20} $-x^{-6} + 5x^{-4} - 9x^{-2} + 11 - 9x^2 + 5x^4 - x^6$

9_{21} $-2x^{-4} + 11x^{-2} - 17 + 11x^2 - 2x^4$

9_{22} $x^{-6} - 5x^{-4} + 10x^{-2} - 11 + 10x^2 - 5x^4 + x^6$

9_{23} $4x^{-4} - 11x^{-2} + 15 - 11x^2 + 4x^4$

9_{24} $-x^{-6} + 5x^{-4} - 10x^{-2} + 13 - 10x^2 + 5x^4 - x^6$

9_{25} $-3x^{-4} + 12x^{-2} - 17 + 12x^2 - 3x^4$

9_{26} $x^{-6} - 5x^{-4} + 11x^{-2} - 13 + 11x^2 - 5x^4 + x^6$

9_{27} $-x^{-6} + 5x^{-4} - 11x^{-2} + 15 - 11x^2 + 5x^4 - x^6$

9_{28} $x^{-6} - 5x^{-4} + 12x^{-2} - 15 + 12x^2 - 5x^4 + x^6$

9_{29} $x^{-6} - 5x^{-4} + 12x^{-2} - 15 + 12x^2 - 5x^4 + x^6$

9_{30} $-x^{-6} + 5x^{-4} - 12x^{-2} + 17 - 12x^2 + 5x^4 - x^6$

9_{31} $x^{-6} - 5x^{-4} + 13x^{-2} - 17 + 13x^2 - 5x^4 + x^6$

9_{32} $x^{-6} - 6x^{-4} + 14x^{-2} - 17 + 14x^2 - 6x^4 + x^6$

9_{33} $-x^{-6} + 6x^{-4} - 14x^{-2} + 19 - 14x^2 + 6x^4 - x^6$

9_{34} $-x^{-6} + 6x^{-4} - 16x^{-2} + 23 - 16x^2 + 6x^4 - x^6$

9_{35} $7x^{-2} - 13 + 7x^2$

9_{36} $-x^{-6} + 5x^{-4} - 8x^{-2} + 9 - 8x^2 + 5x^4 - x^6$

9_{37} $2x^{-4} - 11x^{-2} + 19 - 11x^2 + 2x^4$

9_{38} $5x^{-4} - 14x^{-2} + 19 - 14x^2 + 5x^4$

9_{39} $-3x^{-4} + 14x^{-2} - 21 + 14x^2 - 3x^4$

9_{40} $x^{-6} - 7x^{-4} + 18x^{-2} - 23 + 18x^2 - 7x^4 + x^6$

9_{41} $3x^{-4} - 12x^{-2} + 19 - 12x^2 + 3x^4$

9_{42} $-x^{-4} + 2x^{-2} - 1 + 2x^2 - x^4$

9_{43} $-x^{-6} + 3x^{-4} - 2x^{-2} + 1 - 2x^2 + 3x^4 - x^6$

9_{44} $x^{-4} - 4x^{-2} + 7 - 4x^2 + x^4$

9_{45} $-x^{-4} + 6x^{-2} - 9 + 6x^2 - x^4$

9_{46} $-2x^{-2} + 5 - 2x^2$

9_{47} $x^{-6} - 4x^{-4} + 6x^{-2} - 5 + 6x^2 - 4x^4 + x^6$

9_{48} $-x^{-4} + 7x^{-2} - 11 + 7x^2 - x^4$

9_{49} $3x^{-4} - 6x^{-2} + 7 - 6x^2 + 3x^4$

Conway polynomial

3_1 $1 + z^2$

4_1 $1 - z^2$

5_1 $1 + 3z^2 + z^4$

5_2 $1 + 2z^2$

6_1 $1 - 2z^2$

6_2 $1 - z^2 - z^4$

6_3 $1 + z^2 + z^4$

7_1 $1 + 6z^2 + 5z^4 + z^6$

7_2 $1 + 3z^2$

7_3 $1 + 5z^2 + 2z^4$

7_4 $1 + 4z^2$

7_5 $1 + 4z^2 + 2z^4$

7_6 $1 + z^2 - z^4$

7_7 $1 - z^2 + z^4$

8_1 $1 - 3z^2$

8_2 $1 - 3z^4 - z^6$

8_3 $1 - 4z^2$

8_4 $1 - 3z^2 - 2z^4$

8_5 $1 - z^2 - 3z^4 - z^6$

8_6 $1 - 2z^2 - 2z^4$

8_7 $1 + 2z^2 + 3z^4 + z^6$

8_8 $1 + 2z^2 + 2z^4$

8_9 $1 - 2z^2 - 3z^4 - z^6$

8_{10} $1 + 3z^2 + 3z^4 + z^6$

8_{11} $1 - z^2 - 2z^4$

8_{12} $1 - 3z^2 + z^4$

8_{13} $1 + z^2 + 2z^4$

8_{14} $1 - 2z^4$

8_{15} $1 + 4z^2 + 3z^4$

8_{16} $1 + z^2 + 2z^4 + z^6$

8_{17} $1 - z^2 - 2z^4 - z^6$

8_{18} $1 + z^2 - z^4 - z^6$

8_{19} $1 + 5z^2 + 5z^4 + z^6$

8_{20} $1 + 2z^2 + z^4$

8_{21} $1 - z^4$

9_1 $1 + 10z^2 + 15z^4 + 7z^6 + z^8$

9_2 $1 + 4z^2$

9_3 $1 + 9z^2 + 9z^4 + 2z^6$

9_4 $1 + 7z^2 + 3z^4$

9_5 $1 + 6z^2$

9_6 $1 + 7z^2 + 8z^4 + 2z^6$

9_7 $1 + 5z^2 + 3z^4$

9_8 $1 - 2z^4$

9_9 $1 + 8z^2 + 8z^4 + 2z^6$

9_{10} $1 + 8z^2 + 4z^4$

9_{11} $1 + 4z^2 - z^4 - z^6$

9_{12} $1 + z^2 - 2z^4$

9_{13} $1 + 7z^2 + 4z^4$

9_{14} $1 - z^2 + 2z^4$

9_{15} $1 + 2z^2 - 2z^4$

9_{16} $1 + 6z^2 + 7z^4 + 2z^6$

9_{17} $1 - 2z^2 + z^4 + z^6$

9_{18} $1 + 6z^2 + 4z^4$

9_{19} $1 - 2z^2 + 2z^4$

9_{20} $1 + 2z^2 - z^4 - z^6$

9_{21} $1 + 3z^2 - 2z^4$

9_{22} $1 - z^2 + z^4 + z^6$

9_{23} $1 + 5z^2 + 4z^4$

9_{24} $1 + z^2 - z^4 - z^6$

9_{25} $1 - 3z^4$

9_{26} $1 + z^4 + z^6$

9_{27} $1 - z^4 - z^6$

9_{28} $1 + z^2 + z^4 + z^6$

9_{29} $1 + z^2 + z^4 + z^6$

9_{30} $1 - z^2 - z^4 - z^6$

9_{31} $1 + 2z^2 + z^4 + z^6$

9_{32} $1 - z^2 + z^6$

9_{33} $1 + z^2 - z^6$

9_{34} $1 - z^2 - z^6$

9_{35} $1 + 7z^2$

9_{36} $1 + 3z^2 - z^4 - z^6$

9_{37} $1 - 3z^2 + 2z^4$

9_{38} $1 + 6z^2 + 5z^4$

9_{39} $1 + 2z^2 - 3z^4$

9_{40} $1 - z^2 - z^4 + z^6$

9_{41} $1 + 3z^4$

9_{42} $1 - 2z^2 - z^4$

9_{43} $1 + z^2 - 3z^4 - z^6$

9_{44} $1 + z^4$

9_{45} $1 + 2z^2 - z^4$

9_{46} $1 - 2z^2$

9_{47} $1 - z^2 + 2z^4 + z^6$

9_{48} $1 + 3z^2 - z^4$

9_{49} $1 + 6z^2 + 3z^4$

Jones polynomial

3_1 $-t^{-4} + t^{-3} + t^{-1}$

4_1 $t^{-2} - t^{-1} + 1 - t + t^2$

5_1 $-t^{-7} + t^{-6} - t^{-5} + t^{-4} + t^{-2}$

5_2 $-t^{-6} + t^{-5} - t^{-4} + 2t^{-3} - t^{-2} + t^{-1}$

6_1 $t^{-4} - t^{-3} + t^{-2} - 2t^{-1} + 2 - t + t^2$

6_2 $t^{-5} - 2t^{-4} + 2t^{-3} - 2t^{-2} + 2t^{-1} - 1 + t$

6_3 $-t^{-3} + 2t^{-2} - 2t^{-1} + 3 - 2t + 2t^2 - t^3$

7_1 $-t^{-10} + t^{-9} - t^{-8} + t^{-7} - t^{-6} + t^{-5} + t^{-3}$

7_2 $-t^{-8} + t^{-7} - t^{-6} + 2t^{-5} - 2t^{-4} + 2t^{-3} - t^{-2} + t^{-1}$

7_3 $t^2 - t^3 + 2t^4 - 2t^5 + 3t^6 - 2t^7 + t^8 - t^9$

7_4 $t - 2t^2 + 3t^3 - 2t^4 + 3t^5 - 2t^6 + t^7 - t^8$

7_5 $-t^{-9} + 2t^{-8} - 3t^{-7} + 3t^{-6} - 3t^{-5} + 3t^{-4} - t^{-3} + t^{-2}$

7_6 $-t^{-6} + 2t^{-5} - 3t^{-4} + 4t^{-3} - 3t^{-2} + 3t^{-1} - 2 + t$

7_7 $\qquad -t^{-3} + 3t^{-2} - 3t^{-1} + 4 - 4t + 3t^2 - 2t^3 + t^4$

8_1 $\qquad t^{-6} - t^{-5} + t^{-4} - 2t^{-3} + 2t^{-2} - 2t^{-1} + 2 - t + t^2$

8_2 $\qquad t^{-8} - 2t^{-7} + 2t^{-6} - 3t^{-5} + 3t^{-4} - 2t^{-3} + 2t^{-2} - t^{-1} + 1$

8_3 $\qquad t^{-4} - t^{-3} + 2t^{-2} - 3t^{-1} + 3 - 3t + 2t^2 - t^3 + t^4$

8_4 $\qquad t^{-3} - t^{-2} + 2t^{-1} - 3 + 3t - 3t^2 + 3t^3 - 2t^4 + t^5$

8_5 $\qquad 1 - t + 3t^2 - 3t^3 + 3t^4 - 4t^5 + 3t^6 - 2t^7 + t^8$

8_6 $\qquad t^{-7} - 2t^{-6} + 3t^{-5} - 4t^{-4} + 4t^{-3} - 4t^{-2} + 3t^{-1} - 1 + t$

8_7 $\qquad -t^{-2} + 2t^{-1} - 2 + 4t - 4t^2 + 4t^3 - 3t^4 + 2t^5 - t^6$

8_8 $\qquad -t^{-3} + 2t^{-2} - 3t^{-1} + 5 - 4t + 4t^2 - 3t^3 + 2t^4 - t^5$

8_9 $\qquad t^{-4} - 2t^{-3} + 3t^{-2} - 4t^{-1} + 5 - 4t + 3t^2 - 2t^3 + t^4$

8_{10} $\qquad -t^{-2} + 2t^{-1} - 3 + 5t - 4t^2 + 5t^3 - 4t^4 + 2t^5 - t^6$

8_{11} $\qquad t^{-7} - 2t^{-6} + 3t^{-5} - 5t^{-4} + 5t^{-3} - 4t^{-2} + 4t^{-1} - 2 + t$

8_{12} $\qquad t^{-4} - 2t^{-3} + 4t^{-2} - 5t^{-1} + 5 - 5t + 4t^2 - 2t^3 + t^4$

8_{13} $\qquad -t^{-3} + 3t^{-2} - 4t^{-1} + 5 - 5t + 5t^2 - 3t^3 + 2t^4 - t^5$

8_{14} $\qquad t^{-7} - 3t^{-6} + 4t^{-5} - 5t^{-4} + 6t^{-3} - 5t^{-2} + 4t^{-1} - 2 + t$

8_{15} $\qquad t^{-10} - 3t^{-9} + 4t^{-8} - 6t^{-7} + 6t^{-6} - 5t^{-5} + 5t^{-4} - 2t^{-3} + t^{-2}$

8_{16} $\qquad -t^{-6} + 3t^{-5} - 5t^{-4} + 6t^{-3} - 6t^{-2} + 6t^{-1} - 4 + 3t - t^2$

8_{17} $\qquad t^{-4} - 3t^{-3} + 5t^{-2} - 6t^{-1} + 7 - 6t + 5t^2 - 3t^3 + t^4$

8_{18} $\qquad t^{-4} - 4t^{-3} + 6t^{-2} - 7t^{-1} + 9 - 7t + 6t^2 - 4t^3 + t^4$

8_{19} $\qquad t^3 + t^5 - t^8$

8_{20} $\qquad -t^{-5} + t^{-4} - t^{-3} + 2t^{-2} - t^{-1} + 2 - t$

8_{21} $\qquad t^{-7} - 2t^{-6} + 2t^{-5} - 3t^{-4} + 3t^{-3} - 2t^{-2} + 2t^{-1}$

9_1 $\qquad t^4 + t^6 - t^7 + t^8 - t^9 + t^{10} - t^{11} + t^{12} - t^{13}$

9_2 $\qquad t - t^2 + 2t^3 - 2t^4 + 2t^5 - 2t^6 + 2t^7 - t^8 + t^9 - t^{10}$

9_3 $\qquad t^3 - t^4 + 2t^5 - 2t^6 + 3t^7 - 3t^8 + 3t^9 - 2t^{10} + t^{11} - t^{12}$

9_4 $\qquad t^2 - t^3 + 2t^4 - 3t^5 + 4t^6 - 3t^7 + 3t^8 - 2t^9 + t^{10} - t^{11}$

9_5 $\qquad t - 2t^2 + 3t^3 - 3t^4 + 4t^5 - 3t^6 + 3t^7 - 2t^8 + t^9 - t^{10}$

9_6 $\qquad t^3 - t^4 + 3t^5 - 3t^6 + 4t^7 - 5t^8 + 4t^9 - 3t^{10} + 2t^{11} - t^{12}$

9_7 $\qquad t^2 - t^3 + 3t^4 - 4t^5 + 5t^6 - 5t^7 + 4t^8 - 3t^9 + 2t^{10} - t^{11}$

9_8 $\qquad t^{-3} - 2t^{-2} + 3t^{-1} - 4 + 5t - 5t^2 + 5t^3 - 3t^4 + 2t^5 - t^6$

9_9 $\qquad t^3 - t^4 + 3t^5 - 4t^6 + 5t^7 - 5t^8 + 5t^9 - 4t^{10} + 2t^{11} - t^{12}$

9_{10} $\qquad t^2 - 2t^3 + 4t^4 - 5t^5 + 6t^6 - 5t^7 + 5t^8 - 3t^9 + t^{10} - t^{11}$

9_{11} $\qquad 1 - 2t + 3t^2 - 4t^3 + 6t^4 - 5t^5 + 5t^6 - 4t^7 + 2t^8 - t^9$

9_{12} $\qquad t^{-1} - 2 + 4t - 5t^2 + 6t^3 - 6t^4 + 5t^5 - 3t^6 + 2t^7 - t^8$

9_{13} $\qquad t^2 - 2t^3 + 4t^4 - 5t^5 + 7t^6 - 6t^7 + 5t^8 - 4t^9 + 2t^{10} - t^{11}$

9_{14} $\qquad -t^{-3} + 3t^{-2} - 4t^{-1} + 6 - 6t + 6t^2 - 5t^3 + 3t^4 - 2t^5 + t^6$

9_{15} $\qquad t^{-1} - 2 + 4t - 6t^2 + 7t^3 - 6t^4 + 6t^5 - 4t^6 + 2t^7 - t^8$

9_{16} $\qquad t^3 - t^4 + 4t^5 - 5t^6 + 6t^7 - 7t^8 + 6t^9 - 5t^{10} + 3t^{11} - t^{12}$

9_{17} $\qquad t^{-3} - 2t^{-2} + 4t^{-1} - 5 + 6t - 7t^2 + 6t^3 - 4t^4 + 3t^5 - t^6$

9_{18} $t^2 - 2t^3 + 5t^4 - 6t^5 + 7t^6 - 7t^7 + 6t^8 - 4t^9 + 2t^{10} - t^{11}$

9_{19} $t^{-4} - 2t^{-3} + 4t^{-2} - 6t^{-1} + 7 - 7t + 6t^2 - 4t^3 + 3t^4 - t^5$

9_{20} $1 - 2t + 4t^2 - 5t^3 + 7t^4 - 7t^5 + 6t^6 - 5t^7 + 3t^8 - t^9$

9_{21} $t^{-1} - 3 + 5t - 6t^2 + 8t^3 - 7t^4 + 6t^5 - 4t^6 + 2t^7 - t^8$

9_{22} $t^{-3} - 2t^{-2} + 4t^{-1} - 6 + 7t - 7t^2 + 7t^3 - 5t^4 + 3t^5 - t^6$

9_{23} $t^2 - 2t^3 + 5t^4 - 6t^5 + 8t^6 - 8t^7 + 6t^8 - 5t^9 + 3t^{10} - t^{11}$

9_{24} $t^{-4} - 3t^{-3} + 5t^{-2} - 7t^{-1} + 8 - 7t + 7t^2 - 4t^3 + 2t^4 - t^5$

9_{25} $t^{-1} - 2 + 5t - 7t^2 + 8t^3 - 8t^4 + 7t^5 - 5t^6 + 3t^7 - t^8$

9_{26} $-t^{-2} + 3t^{-1} - 4 + 7t - 8t^2 + 8t^3 - 7t^4 + 5t^5 - 3t^6 + t^7$

9_{27} $t^{-4} - 3t^{-3} + 5t^{-2} - 7t^{-1} + 9 - 8t + 7t^2 - 5t^3 + 3t^4 - t^5$

9_{28} $-t^{-2} + 3t^{-1} - 5 + 8t - 8t^2 + 9t^3 - 8t^4 + 5t^5 - 3t^6 + t^7$

9_{29} $t^{-3} - 3t^{-2} + 5t^{-1} - 7 + 9t - 8t^2 + 8t^3 - 6t^4 + 3t^5 - t^6$

9_{30} $t^{-4} - 3t^{-3} + 6t^{-2} - 8t^{-1} + 9 - 9t + 8t^2 - 5t^3 + 3t^4 - t^5$

9_{31} $-t^{-2} + 3t^{-1} - 5 + 8t - 9t^2 + 10t^3 - 8t^4 + 6t^5 - 4t^6 + t^7$

9_{32} $-t^{-2} + 4t^{-1} - 6 + 9t - 10t^2 + 10t^3 - 9t^4 + 6t^5 - 3t^6 + t^7$

9_{33} $t^{-4} - 4t^{-3} + 7t^{-2} - 9t^{-1} + 11 - 10t + 9t^2 - 6t^3 + 3t^4 - t^5$

9_{34} $t^{-4} - 4t^{-3} + 8t^{-2} - 10t^{-1} + 12 - 12t + 10t^2 - 7t^3 + 4t^4 - t^5$

9_{35} $t - 2t^2 + 3t^3 - 4t^4 + 5t^5 - 3t^6 + 4t^7 - 3t^8 + t^9 - t^{10}$

9_{36} $1 - 2t + 4t^2 - 5t^3 + 6t^4 - 6t^5 + 6t^6 - 4t^7 + 2t^8 - t^9$

9_{37} $t^{-4} - 2t^{-3} + 5t^{-2} - 7t^{-1} + 7 - 8t + 7t^2 - 4t^3 + 3t^4 - t^5$

9_{38} $t^2 - 3t^3 + 7t^4 - 8t^5 + 10t^6 - 10t^7 + 8t^8 - 6t^9 + 3t^{10} - t^{11}$

9_{39} $t^{-1} - 3 + 6t - 8t^2 + 10t^3 - 9t^4 + 8t^5 - 6t^6 + 3t^7 - t^8$

9_{40} $-t^{-2} + 5t^{-1} - 8 + 11t - 13t^2 + 13t^3 - 11t^4 + 8t^5 - 4t^6 + t^7$

9_{41} $-t^{-3} + 3t^{-2} - 5t^{-1} + 8 - 8t + 8t^2 - 7t^3 + 5t^4 - 3t^5 + t^6$

9_{42} $t^{-3} - t^{-2} + t^{-1} - 1 + t - t^2 + t^3$

9_{43} $1 - t + 2t^2 - 2t^3 + 2t^4 - 2t^5 + 2t^6 - t^7$

9_{44} $t^{-2} - 2t^{-1} + 3 - 3t + 3t^2 - 2t^3 + 2t^4 - t^5$

9_{45} $2t - 3t^2 + 4t^3 - 4t^4 + 4t^5 - 3t^6 + 2t^7 - t^8$

9_{46} $2 - t + t^2 - 2t^3 + t^4 - t^5 + t^6$

9_{47} $-t^{-2} + 3t^{-1} - 3 + 5t - 5t^2 + 4t^3 - 4t^4 + 2t^5$

9_{48} $t^{-1} - 3 + 4t - 4t^2 + 6t^3 - 4t^4 + 3t^5 - 2t^6$

9_{49} $t^2 - 2t^3 + 4t^4 - 4t^5 + 5t^6 - 4t^7 + 3t^8 - 2t^9$

Appendix E Polygon Coordinates

These are the coordinates of the polygons shown in Figure 1.6.

0_1 $(4, 9, 5)$, $(7, -9, 5)$, $(-9, -3, 5)$.

3_1 $(4, 9, 5)$, $(-7, -7, -5)$, $(7, -9, 5)$, $(-1, 9, -5)$, $(-9, -3, 5)$, $(9, -5, -5)$.

4_1 $(9, -6, 3)$, $(-4, -7, 3)$, $(1, 7, 2)$, $(-9, 2, -10)$, $(4, -5, 10)$, $(2, 2, -2)$, $(-5, 2, 5)$.

5_1 $(-4, -6, 0)$, $(-1, -8, 9)$, $(6, 1, -8)$, $(-7, 1, 0)$, $(3, -8, 3)$, $(7, -2, 6)$, $(0, -1, -8)$, $(-5, 8, -1)$.

5_2 $(3, 6, 2)$, $(9, 2, -4)$, $(1, -1, 7)$, $(0, -3, -5)$, $(-8, -4, 1)$, $(10, 7, 1)$, $(-9, 8, -6)$, $(2, -8, 3)$.

6_1 $(0, -8, -5)$, $(-3, -7, 3)$, $(8, 10, -4)$, $(-9, 0, 5)$, $(8, -10, -5)$, $(-3, 7, 6)$, $(0, 8, -5)$, $(5, 0, 5)$.

6_2 $(-3, -10, 7)$, $(-7, -8, -10)$, $(-2, -1, 10)$, $(5, 4, -7)$, $(-5, 10, -2)$, $(5, -10, 10)$, $(-10, -4, -6)$, $(10, 9, -1)$.

6_3 $(2, -8, -8)$, $(-3, -2, -1)$, $(1, 3, -10)$, $(-1, 9, 19)$, $(-4, 1, -7)$, $(7, 9, -2)$, $(-6, -7, -4)$, $(-1, 5, -6)$.

8_{19} $(-4, 6, 5)$, $(-4, -5, 3)$, $(1, 2, 3)$, $(-3, 1, 2)$, $(5, -3, 7)$, $(1, 6, -10)$, $(-3, -2, 10)$, $(5, -5, 0)$.

8_{20} $(-3, 3, -3)$, $(4, -5, -1)$, $(0, 5, -3)$, $(-5, -5, 6)$, $(5, 1, -6)$, $(0, 2, 1)$, $(-2, -2, -5)$, $(6, 3, -1)$.

$3_1 \# 3_1$ $(-6.0, -1.4, 0.8)$, $(5.1, -2.8, -4.9)$, $(1.9, 1.4, 4.6)$, $(-0.3, -3.2, -6.7)$, $(4.3, -3.6, 3.3)$, $(-2.7, 4.6, -3.6)$, $(-3.0, -6.0, -0.8)$, $(2.6, 4.9, -1.5)$.

$3_1 \# 3_1^*$ $(-3.0, -5.0, -1.7)$, $(8.0, 5.0, 1.3)$, $(-8.0, 5.0, -1.1)$, $(3.0, -5.0, 0.9)$, $(-1.0, 3.5, -3.7)$, $(-1.5, 0.5, 15)$, $(1.5, 0.5, -15)$, $(1.0, 3.5, 3.6)$.

Appendix F   Family Properties

Torus knots

Let K be the (p, q) torus knot with $p > q \geqslant 2$.

Torus knots are reversible and cheiral. Torus knots are positive. Torus knots are alternating for $q = 2$ and non-alternating for $q > 2$.

$$
\begin{aligned}
c(K) &= p(q-1) & \text{Proposition 10.5.3} \\
p(K) &\leqslant 2p & \S1.5 \\
g(K) &= \tfrac{1}{2}(p-1)(q-1) & \text{Corollary 7.6.7} \\
u(K) &= \tfrac{1}{2}(p-1)(q-1) & [247, 248] \\
b(K) &= q & [108] \\
s(K) &= q & \text{Proposition 10.5.2} \\
\alpha(K) &= p + q & \text{Figure 10.11}
\end{aligned}
$$

Rational links

Let L be the p/q rational link with q odd, $p > 0$ and $-p < q < p$.

Let $C(a_1, a_2, \ldots, a_{n-1}, a_n)$ be the continued fraction expansion of p/q with all $a_i > 0$.

Let $C(b_1, b_2, \ldots, b_{m-1}, b_m)$ be the continued fraction expansion of p/q with all b_i even, and let t be the number of times there is a change of sign between adjacent coefficients b_i and b_{i+1}.

Rational links are reversible and alternating.

$$
\begin{aligned}
c(L) &= \sum a_i & \text{Theorem 8.7.1 and Corollary 9.5.10} \\
p(L) &\leqslant c(L) + 3 & [250] \\
g(L) &= \tfrac{1}{2}(m - \mu(L) + 1) & \text{Corollary 8.7.5} \\
\det(L) &= p & \text{Theorem 8.7.7} \\
u(L) & & \text{partial result in } [246] \\
b(L) &= 2 & \text{Theorem 4.10.3} \\
s(L) &= \sum |b_i| - t + 1 & \S10.4 \\
\alpha(L) &= c(L) + 2 & [224]
\end{aligned}
$$

Bibliography

General

Books and survey articles ordered by year of publication. (The other sections are ordered alphabetically by author.)

1. K. REIDEMEISTER, *Knoten Theorie*, Springer-Verlag, 1932. English translation: *Knot Theory*, BCS Associates, 1983.

2. R. H. FOX, 'A quick trip through knot theory', *Topology of 3-Manifolds and Related Topics*, ed. M. K. Fort, Prentice-Hall Inc., 1962, pp. 120–167.

3. R. H. CROWELL and R. H. FOX, *Introduction to Knot Theory*, Ginn and Co., 1963.

4. J. S. BIRMAN, *Braids, Links and Mapping Class Groups*, Annals of Math. Studies **84**, Princeton Univ. Press, 1974.

5. D. ROLFSEN, *Knots and Links*, Publish or Perish Inc., 1976.

6. C. McA. GORDON, 'Some aspects of classical knot theory', *Knot Theory* (Proc. Conference Plans-sur-Bex 1977), ed. J. C. Hausmann, Lecture Notes in Math. **685**, Springer-Verlag, 1978, pp. 1–60.

7. L. H. KAUFFMAN, *Formal Knot Theory*, Princeton Univ. Press, 1983.

8. S. MORAN, *The Mathematical Theory of Knots and Braids: An Introduction*, Math. Studies **82**, North-Holland, 1983.

9. G. BURDE and H. ZIESCHANG, *Knots*, de Gruyter, 1985.

10. L. H. KAUFFMAN, *On Knots*, Annals of Math. Studies **115**, Princeton Univ. Press, 1987.

11. L. H. KAUFFMAN, *Knots and Physics*, Series on Knots and Everything **1**, World Scientific, 1991.

12. G. HEMION, *The Classification of Knots and 3-Dimensional Spaces*, Oxford Univ. Press, 1992.

13. C. LIVINGSTON, *Formal Knot Theory*, Carus Math. Monographs **24**, Math. Association of America, 1993.

14. C. C. ADAMS, *The Knot Book: An Elementary Introduction to the Mathematical Theory of Knots*, W. H. Freeman and Co., 1994.

15. N. D. GILBERT and T. PORTER, *Knots and Surfaces*, Oxford Univ. Press, 1994.

16. P. VAN DE GRIEND, 'A history of topological knot theory', *History and Science of Knots*, eds. J. C. Turner and P. van de Griend, Series on Knots and Everything **11**, World Scientific, 1996, pp. 205–259.

17. K. MURASUGI, *Knot Theory and Its Applications* (translated from the Japanese by Bohdan Kurpita), Birkhäuser, 1996.

18. A. KAWAUCHI, *A Survey of Knot Theory*, Birkhäuser, 1997.

19. W. B. R. LICKORISH, *An Introduction to Knot Theory*, Graduate Texts in Math. **175**, Springer-Verlag, 1997.

20. M. EPPLE, 'Geometric aspects in the development of knot theory', *History of Topology*, ed. I. M. James, Elsevier, 1999, pp. 301–357.

21. E. FLAPAN, *When Topology Meets Chemistry: A Topological Look at Molecular Chirality*, Math. Association of America, Cambridge Univ. Press, 2000.

1 Introduction

22. C. C. ADAMS, B. M. BRENNAN, D. L. GREILSHEIMER and A. K. WOO, 'Stick numbers and the composition of knots and links', *J. Knot Theory and Its Ramifications* **6** (1997) 149–161.

23. J. W. ALEXANDER and G. B. BRIGGS, 'On types of knotted curves', *Annals of Math.* **28** (1927) 562–586.

24. M. G. V. BOGLE, J. E. HEARST, V. F. R. JONES and L. STOILOV, 'Lissajous knots', *J. Knot Theory and Its Ramifications* **3** (1994) 121–140.

25. R. BRODE, *Über wilde Knoten und ihre Anzahl*, Diplomarbeit, Ruhr-Univ. Bochum, 1981.

26. A. CAUDRON, *Classification des noeuds et des enlacements*, Publications Math. d'Orsay **82**, Univ. Paris Sud, 1982.

27. C. CERF, 'Atlas of oriented knots and links', *Topology Atlas* **3** no 2 (1998) 1–32.

28. C. CERF, 'The topological chirality of knots and links', *Chemical Topology: Applications and Techniques*, eds. D. Bonchev and D. H. Rouvray, Math. Chemistry Series **6**, Gordon and Breach Science Publ., 2000, pp. 1–34.

29. P. R. CROMWELL, E. BELTRAMI and M. RAMPICHINI, 'The Borromean rings', *Math. Intelligencer* **20** no 1 (1998) 53–62.

30. M. DEHN, 'Die beiden Kleeblattschlingen', *Math. Annalen* **75** (1914) 402–413.

31. H. DOLL and J. HOSTE, 'A tabulation of oriented links', *Math. Computation* **57** (1991) 747–761. Diagrams are in Appendix A on microfiche pp. A1–A10.

32. C. H. DOWKER and M. B. THISTLETHWAITE, 'On the classification of knots', *C.R. Math. Reports Acad. Science Canada* **IV** (1982) 129–131.

33. C. H. DOWKER and M. B. THISTLETHWAITE, 'Classification of knot projections', *Topology and Its Applications* **16** (1983) 19–31.

34. R. H. FOX, 'A remarkable simple closed curve', *Annals of Math.* **50** (1949) 264–265.

35. R. H. FOX and E. ARTIN, 'Some wild cells and spheres in three-dimensional space', *Annals of Math.* **49** (1948) 979–990.

36. J. HOSTE, M. B. THISTLETHWAITE and J. WEEKS, 'The first 1 701 936 knots', *Math. Intelligencer* **20** no 4 (1998) 33–48.

37. B. JIANG, X.-S. LIN, S. WANG and Y.-Q. WU, 'Achirality of knots and links', *Topology and Its Applications* **119** (2002) 185–208.

38. G. T. JIN, 'Polygon indices and superbridge indices of torus knots and links', *J. Knot Theory and Its Ramifications* **6** (1997) 281–289.

39. G. T. JIN and H. S. KIM, 'Polygonal knots', *J. Korean Math. Soc.* **30** (1993) 371–383.

40. V. F. R. JONES and J. H. PRZYTYCKI, 'Lissajous knots and billiard knots', *Knot Theory* (Proc. Conference Warsaw 1995), eds. V. F. R. Jones *et al.*, Banach Center Publications **42**, 1998, pp. 145–163.

41. L. H. KAUFFMAN, 'Fourier knots', *Ideal Knots*, eds. A. Stasiak, V. Katritch and L. H. Kauffman, Series on Knots and Everything **19**, World Scientific, 1998, pp. 364–373.

42. T. P. KIRKMAN, 'The enumeration, description and construction of knots of fewer than ten crossings', *Trans. Royal Soc. Edinburgh* **32** (1885) 281–309.

43. T. P. KIRKMAN, 'The 364 unifilar knots of ten crossings enumerated and defined', *Trans. Royal Soc. Edinburgh* **32** (1885) 483–506.

44. C. LAMM, *Lissajous Knoten*, Diplom Thesis, Bonn Univ., 1996.

45. C. LAMM, 'There are infinitely many Lissajous knots', *Manuscripta Math.* **93** (1997) 29–37.

46. C. LIANG, C. CERF and K. MISLOW, 'Specification of chirality for links and knots', *J. Math. Chemistry* **19** (1996) 241–263.

47. C. N. LITTLE, 'On knots, with a census of order ten', *Trans. Connecticut Acad. Science* **18** (1885) 374–378.

48. C. N. LITTLE, 'Nonalternate ± knots of orders eight and nine', *Trans. Royal Soc. Edinburgh* **35** (1889) 663–664.

49. C. N. LITTLE, 'Alternate ± knots of order 11', *Trans. Royal Soc. Edinburgh* **36** (1890) 253–255.

50. C. N. LITTLE, 'Nonalternate ± knots', *Trans. Royal Soc. Edinburgh* **39** (1900) 771–778.

51. M. MEISSEN, *Lower and Upper Bounds on Edge Numbers and Crossing Numbers of Knots*, Ph.D. Thesis, Iowa Univ., 1997.

52. M. MEISSEN, 'Edge number results for piecewise-linear knots', *Knot Theory* (Proc. Conference Warsaw 1995), eds. V. F. R. Jones *et al.*, Banach Center Publications **42**, 1998, pp. 235–242.

53. J. W. MILNOR, 'Most knots are wild', *Fundamenta Math.* **54** (1964) 335–338.

54. R. RANDELL, 'Invariants of piecewise-linear knots', *Knot Theory* (Proc. Conference Warsaw 1995), eds. V. F. R. Jones *et al.*, Banach Center Publications **42**, 1998, pp. 307–319.

55. E. J. RAWDON and R. G. SCHAREIN, 'Upper bounds for equilateral stick numbers', *Physical Knots: Knotting, Linking and Folding Geometric Objects in* $\mathbb{R}^3$ (Proc. Conference Las Vegas 2001), eds. J. A. Calvo, K. C. Millett and E. J. Rawdon, Contemporary Math. **304**, American Math. Soc., 2002, pp. 55–75.

56. P. G. TAIT, 'On knots', *Trans. Royal Soc. Edinburgh* **28** (1876) 145–190.

57. P. G. TAIT, 'On knots II', *Trans. Royal Soc. Edinburgh* **32** (1884) 327–342.

58. P. G. TAIT, 'On knots III', *Trans. Royal Soc. Edinburgh* **32** (1885) 493–506.

59. M. B. THISTLETHWAITE, 'Knot tabulations and related topics', *Aspects of Topology: In Memory of Hugh Dowker 1912–1982*, eds. I. M. James and E. H. Kronheimer, London Math. Soc. Lecture Note Series **93**, Cambridge Univ. Press, 1985, pp. 1–76.

60. W. H. THOMSON, 'On vortex motion', *Trans. Royal Soc. Edinburgh* **25** (1869) 217–260.

61. H. TIETZE, 'Über die topologischen Invarianten mehrdimensionaler Mannigfaltigkeiten', *Monatshefte für Math. und Physik* **19** (1908) 1–118.

62. A. TRAUTWEIN, *Harmonic Knots*, Ph.D. Thesis, Iowa Univ., 1994.

63. H. F. TROTTER, 'Non-invertible knots exist', *Topology* **2** (1964) 341–358.

64. H. WENDT, 'Die gordische Auflösung von Knoten', *Math. Zeitschrift* **42** (1937) 680–696.

2   A Topologist's Toolkit

65. J. W. ALEXANDER, 'An example of a simply connected surface bounding a region which is not simply connected', *Proc. National Acad. Sciences USA* **10** (1924) 10–12.

66. R. H. BING, *The Geometric Topology of 3-Manifolds*, Colloquium Publications **40**, American Math. Soc., 1983.

67. B. BOLLOBÁS, *Graph Theory: An Introductory Course*, Graduate Texts in Math. **63**, Springer-Verlag, 1979.

68. J. S. CARTER, *How Surfaces Intersect in Space: An Introduction to Topology*, Series on Knots and Everything **2**, World Scientific, 1993.

69. I. FÁRY, 'On straight lines representation of planar graphs', *Acta Scientiarum Math. Szeged* **11** (1948) 229–233.

70. E. FLAPAN, 'Rigid and non-rigid achirality', *Pacific J. Math.* **129** (1987) 57–66.

71. G. K. FRANCIS and J. R. WEEKS, 'Conway's ZIP proof', *American Math. Monthly* **106** (May, 1999) 393–399.

72. J. HEMPEL, *3-Manifolds*, Princeton Univ. Press, 1976.

73. W. H. JACO, *Lectures on Three-Manifold Topology*, CBMS Regional Conference Series **43**, American Math. Soc., 1980.

74. E. E. MOISE, *Geometric Topology in Dimensions 2 and 3*, Graduate Texts in Math. **47**, Springer-Verlag, 1977.

75. W. D. NEUMANN, 'Notes on geometry and 3-manifolds', *Low Dimensional Topology* (Proc. Conference Budapest 1998), eds. K. Boroczky, W. Neumann, and A. Stipsicz, Bolyai Soc. Math. Studies **8**, 1999, pp. 191–267.

76. C. P. ROURKE and B. J. SANDERSON, *Introduction to Piecewise-Linear Topology*, Springer-Verlag, 1982.

77. S. K. STEIN, 'Convex maps', *Proc. American Math. Soc.* **2** (1951) 464–466.

78. W. A. SUTHERLAND, *Introduction to Metric and Topological Spaces*, Oxford Univ. Press, 1975.

79. K. WAGNER, 'Bemerkungen zum Vierfarbenproblem', *Jahresberichte Deutsch Math.-Vereinigung* **46** (1936) 26–32.

3   Link Diagrams

80. V. I. ARNOLD, 'Plane curves, their invariants, perestroikas and classifications', *Advances in Soviet Math.* **21** (1994) 33–91.

81. V. I. ARNOLD, *Topological Invariants of Plane Curves and Caustics*, Univ. Lecture Series **5**, American Math. Soc., 1994.

82. S. A. BLEILER, 'A note on unknotting number', *Math. Proc. Cambridge Philos. Soc.* **96** (1984) 469–471.

83. C. MCA. GORDON and J. LUECKE, 'Knots are determined by their complements', *J. American Math. Soc.* **2** (1989) 371–415.

84. W. HAKEN, 'Über Flächen in 3-dimensionalen Mannigfaltigkeiten Lösung des Isotopieproblems für den Kreisknoten', *Proc. International Congress Math.* (Amsterdam 1954) **1** (1957) pp. 481–482.

85. W. HAKEN, 'Theorie der Normalflächen', *Acta Math.* **105** (1961) 245–375.

86. W. HAKEN, 'Über das Homöomorphieproblem der 3-Mannigfaltigkeiten I', *Math. Zeitschrift* **80** (1962) 89–120.

87. J. HASS and J. C. LAGARIAS, 'The number of Reidemeister moves needed for unknotting', *J. American Math. Soc.* **14** (2001) 399–428.

88. J. HASS, J. C. LAGARIAS and N. PIPPENGER, 'The computational complexity of knot and link problems', *J. Assoc. Computing Machinery* **46** (1999) 185–211.

89. G. HEMION, 'On the classification of homeomorphisms of 2-manifolds and the classification of 3-manifolds', *Acta Math.* **142** (1979) 123–155.

90. W. W. MENASCO and M. B. THISTLETHWAITE, 'The Tait flyping conjecture', *Bull. American Math. Soc.* **25** (1991) 403–412.

91. W. W. MENASCO and M. B. THISTLETHWAITE, 'The classification of alternating links', *Annals of Math.* **138** (1993) 113–171.

92. T. NAKAMURA, 'On a positive knot without positive minimal diagrams', *Interdisciplinary Information Sciences* **9** (2003) 61–75.

93. Y. NAKANISHI, 'Unknotting numbers and knot diagrams with the minimum crossings', *Math. Seminar Notes Kobe Univ.* **11** (1983) 257–258.

94. S. NEGAMI, 'Ramsey theorems for knots, links and spatial graphs', *Trans. American Math. Soc.* **324** (1991) 527–541.

95. O.-P. OSTLUND, 'Invariants of knot diagrams and relations among Reidemeister moves', *J. Knot Theory and Its Ramifications* **10** (2001) 1215–1227.

96. K. PERKO, 'On the classification of knots', *Proc. American Math. Soc.* **45** (1974) 262–266.

97. A. STOIMENOW, 'On the crossing number of positive knots and braids and braid index criteria of Jones and Morton–Williams–Franks', *Trans. American Math. Soc.* **354** (2002) 3927–3954.

98. B. TRACE, 'On the Reidemeister moves of a classical knot', *Proc. American Math. Soc.* **89** (1983) 722–724.

99. A. M. TURING, 'Solvable and unsolvable problems', *Science News* **31** (Feb 1954) 7–33.

100. F. WALDHAUSEN, 'Recent results on sufficiently large 3-manifolds', *Algebraic and Geometric Topology* (Proc. Conference Stanford 1976), ed. R. J. Milgram, Proc. Symposium Pure Math. **32** no. 2, American Math. Soc., 1978, pp. 21–38.

4 Constructions and Decompositions of Links

101. H. BRUNN, 'Über Verkettung', *Sitzungsberichte der Bayerische Akad. der Wissenschaften, Math.–Phys. Klasse* **22** (1892) 77–99.

102. J.-W. CHUNG and X.-S. LIN, *On the Bridge Number of Knot Diagrams with Minimal Crossings*, Preprint, Univ. California Riverside, 2003. e-Print archive: `math.GT/0301320`.

103. Y. DIAO, N. PIPPENGER and D. W. SUMNERS, 'On random knots', *J. Knot Theory and Its Ramifications* **3** (1994) 419–429.

104. D. JUNGREIS, 'Gaussian random polygons are globally knotted', *J. Knot Theory and Its Ramifications* **3** (1994) 455–464.

105. M. SCHARLEMANN, 'Unknotting number one knots are prime', *Inventiones Math.* **82** (1985) 37–55.

106. H. SCHUBERT, 'Die eindeutige Zerlagbarkeit eines Knoten in Primknoten', *Sitzungsberichte der Heidelberger Akad. der Wissenschaften, Math.–Nat. Klasse* **3** (1949) 57–104.

107. H. SCHUBERT, 'Knoten und Vollringe', *Acta Math.* **90** (1953) 131–286.

108. H. SCHUBERT, 'Über eine numerische Knoteninvariante', *Math. Zeitschrift* **61** (1954) 245–288.

109. J. SCHULTENS, 'Additivity of bridge numbers of knots', *On Heegaard Splittings and Dehn Surgeries of 3-Manifolds, and Topics Related to Them* (Proc. Conference Kyoto 2001), Surikaisekikenkyusho Kokyuroku **1229**, 2001, pp. 103–115.

110. W. P. THURSTON, 'Three-dimensional manifolds, Kleinian groups and hyperbolic geometry', *Bull. American Math. Soc.* **6** (1982) 357–381.

111. W. P. THURSTON, 'Hyperbolic structures on 3-manifolds I: deformations of acylindrical manifolds', *Annals of Math.* **124** (1986) 203–246.

112. W. P. THURSTON, *Three-Dimensional Geometry and Topology*, Princeton Univ. Press, 1997.

113. J. H. C. WHITEHEAD, 'On doubled knots', *J. London Math. Soc.* **12** (1937) 63–71.

5   Spanning Surfaces and Genus

114. D. BAR-NATAN, J. FULMAN and L. H. KAUFFMAN, 'An elementary proof that all spanning surfaces of a link are tube-equivalent', *J. Knot Theory and Its Ramifications* **7** (1998) 873–879.

115. B. E. CLARK, 'Crosscaps and knots', *International J. Math. and Math. Science* **1** (1978) 113–123.

116. F. FRANKL and L. PONTRJAGIN, 'Ein Knotensatz mit Anwendung auf die Dimensionstheorie', *Math. Annalen* **102** (1930) 785–789.

117. M. HIRASAWA, 'The flat genus of links', *Kobe J. Math.* **12** no 2 (1995) 155–159.

118. A. KAWAUCHI, 'On coefficient polynomials of the skein polynomial of an oriented link', *Kobe J. Math.* **11** no 1 (1994) 49–68.

119. M. KOBAYASHI and T. KOBAYASHI, 'On canonical genus and free genus of a knot', *J. Knot Theory and Its Ramifications* **5** (1996) 77–85.

120. H. C. LYON, 'Simple knots without unique minimal surfaces', *Proc. American Math. Soc.* **43** (1974) 449–454.

121. H. MURAKAMI and A. YASUHARA, 'Crosscap number of a knot', *Pacific J. Math.* **171** (1995) 261–273.

122. H. SEIFERT, 'Über das Geschlecht von Knoten', *Math. Annalen* **110** (1934) 571–592.

123. A. STOIMENOW, *Knots of Genus Two*, Preprint, Ludwig-Maximilians Univ., Munich, 2000. e-Print archive: `math.GT/0303012`.

124. A. STOIMENOW, 'Knots of genus one', *Proc. American Math. Soc.* **129** (2001) 2141–2156.

6 Matrix Invariants

125. D. ERLE, 'Die quadratische Form eines Knotens und ein Satz über Knoten-mannigfaltigkeiten', *J. Reine und Angewandte Math.* **236** (1969) 174–218.

126. P. J. GIBLIN, *Graphs, Surfaces and Homology: An Introduction to Algebraic Topology*, Chapman and Hall, 1977.

127. C. McA. GORDON and R. A. LITHERLAND, 'On the signature of a link', *Inventiones Math.* **47** (1978) 53–69.

128. B. W. JONES, *The Arithmetic Theory of Quadratic Forms*, Carus Math. Monographs, 1950.

129. C. KOSNIOWSKI, *A First Course in Algebraic Topology*, Cambridge Univ. Press, 1980.

130. J. W. MILNOR, 'Infinite cyclic coverings', *Topology of Manifolds* (Proc. Conference Michigan 1967), ed. J. G. Hocking, Prindle, Weber and Schmidt, 1968, pp. 115–133.

131. J. R. MUNKRES, *Elements of Algebraic Topology*, Addison-Wesley, 1984.

132. K. MURASUGI, 'On a certain numerical invariant of link types', *Trans. American Math. Soc.* **117** (1965) 387–422.

133. K. MURASUGI, 'On the signature of links', *Topology* **9** (1970) 283–298.

134. H. POINCARÉ, 'Analysis situs', *J. École Polytechnique* **1** (1895) 1–121.

135. H. F. TROTTER, 'Homology of group systems with applications to knot theory', *Annals of Math.* **76** (1962) 464–498.

7 The Alexander–Conway Polynomial

136. J. W. ALEXANDER, 'Topological invariants for knots and links', *Trans. American Math. Soc.* **30** (1928) 275–306.

137. E. ARTIN, 'Theorie der Zöpfe', *Abhandlungen Math. Seminar Univ. Hamburg* **4** (1925) 47–72.

138. E. ARTIN, 'Theory of braids', *Annals of Math.* **48** (1947) 101–126.

139. J. H. CONWAY, 'An enumeration of knots and links, and some of their algebraic properties', *Computational Problems in Abstract Algebra* (Proc. Conference Oxford 1970), ed. J. Leech, Pergamon Press, 1967, pp. 329–358.

140. P. R. CROMWELL, 'Homogeneous links', *J. London Math. Soc.* **39** (1989) 535–552.

141. P. R. CROMWELL, 'Some infinite families of satellite knots with given Alexander polynomial', *Mathematika* **38** (1991) 156–169.

142. D. GABAI, 'Genera of the arborescent links', *Memoirs of the American Math. Soc.* **59** no. 339, 1986.

143. C. GILLER, 'A family of links and the Conway calculus', *Trans. American Math. Soc.* **270** (1982) 75–109.

144. R. HARTLEY, 'The Conway potential function for links', *Commentarii Math. Helvetici* **58** (1983) 365–378.

145. L. H. KAUFFMAN, 'The Conway polynomial', *Topology* **20** (1981) 101–108.

146. S. KINOSHITA and H. TERASAKA, 'On unions of knots', *Osaka Math. J.* **9** (1957) 131–153.

147. W. B. R. LICKORISH, 'Linear skein theory and link polynomials', *Topology and Its Applications* **27** (1987) 265–274.

148. W. B. R. LICKORISH, 'Polynomials for links', *Bull. London Math. Soc.* **20** (1988) 558–588.

149. H. R. MORTON, 'Polynomials from braids', *Braids* (Proc. Conference Santa Cruz 1986), eds. J. S. Birman and A. Libgober, Contemporary Math. **78**, American Math. Soc., 1988, pp. 575–585.

150. H. R. MORTON and H. B. SHORT, 'Calculating the 2-variable polynomial for knots presented as closed braids', *J. Algorithms* **11** (1990) 117–131.

151. J. H. PRZYTYCKI, 'Classical roots of knot theory', *Chaos, Solitons and Fractals* **9** no. 4–5 (1998) 531–545.

152. R. RILEY, 'Homomorphisms of knot groups on finite groups', *Math. Computation* **25** (1971) 603–619.

8 Rational Tangles

153. C. BANKWITZ and H. G. SCHUMANN, 'Über Viergeflechte', *Abhandlungen Math. Seminar Univ. Hamburg* **10** (1934) 263–284.

154. E. J. BRODY, 'The topological classification of the lens spaces', *Annals of Math.* **71** (1960) 163–184.

155. J. H. CONWAY, see [139].

156. C. ERNST and D. W. SUMNERS, 'The growth of the number of prime knots', *Math. Proc. Cambridge Philos. Soc.* **102** (1987) 303–315.

157. C. ERNST and D. W. SUMNERS, 'A calculus for rational tangles: applications to DNA recombination', *Math. Proc. Cambridge Philos. Soc.* **108** (1990) 489–515.

158. L. GOERITZ, 'Bemerkungen zur Knotentheorie', *Abhandlungen Math. Seminar Univ. Hamburg* **10** (1934) 201–210.

159. J. R. GOLDMAN and L. H. KAUFFMAN, 'Rational tangles', *Advances in Applied Math.* **18** (1997) 300–332.

160. L. H. KAUFFMAN and S. LAMBROPOULOU, 'Classifying and applying rational knots and rational tangles', *Physical Knots: Knotting, Linking and Folding Geometric Objects in* $\mathbb{R}^3$ (Proc. Conference Las Vegas 2001), eds. J. A. Calvo, K. C. Millett and E. J. Rawdon, Contemporary Math. **304**, American Math. Soc., 2002, pp. 223–259.

161. L. H. KAUFFMAN and S. LAMBROPOULOU, 'On the classification of rational tangles', *Advances in Applied Math.*, to appear 2004.

162. L. H. KAUFFMAN and S. LAMBROPOULOU, 'On the classification of rational knots', *L'Enseignement Math.* **49** (2003) 357–410.

163. H. SCHUBERT, 'Knoten mit zwei Brücken', *Math. Zeitschrift* **65** (1956) 133–170.

164. L. SIEBENMANN, *Exercises sur les noeuds rationnels*, unpublished notes, Orsay, 1975.

165. D. W. SUMNERS (ed.), *New Scientific Applications of Geometry and Topology*, Proc. Symposia Applied Math. **45**, American Math. Soc., 1993.

166. D. W. SUMNERS, 'Lifting the curtain: using topology to probe the hidden action of enzymes', *Notices American Math. Soc.* **42** (1995) 528–537.

167. D. W. SUMNERS, C. ERNST, S. SPENGLER and N. COZZARELLI, 'Analysis of the mechanisms of DNA recombination using tangles', *Quarterly Review of Biophysics* **28** (1995) 253–313.

168. C. SUNDBERG and M. B. THISTLETHWAITE, 'The rate of growth of the number of prime alternating links and tangles', *Pacific J. Math.* **182** (1998) 329–358.

9 More Polynomials

169. Y. BAE and H. R. MORTON, 'The spread and extreme terms of Jones polynomials', *J. Knot Theory and Its Ramifications* **12** (2003) 359–373.

170. D. BENNEQUIN, 'Entrelacements et équations de Pfaff', *Astérisque* **107–108** (1983) 87–161.

171. J. S. BIRMAN, 'On the Jones polynomial of closed 3-braids', *Inventiones Math.* **81** (1985) 287–294.

172. J. S. BIRMAN, 'New points of view in knot theory', *Bull. American Math. Soc.* **28** (1993) 253–287.

173. R. D. BRANDT, W. B. R. LICKORISH and K. C. MILLETT, 'A polynomial invariant for unoriented knots and links', *Inventiones Math.* **84** (1986) 563–573.

174. W. BURAU, 'Über Zopfgruppen und gleichsinnig verdrillte Verkettungen', *Abhandlungen Math. Seminar Univ. Hamburg* **11** (1935) 179–186.

175. S. ELIAHOU, L. H. KAUFFMAN and M. B. THISTLETHWAITE, 'Infinite families of links with trivial Jones polynomial', *Topology* **42** (2003) 155–169.

176. P. FREYD and D. YETTER, *A New Invariant for Knots and Links*, Preprint (unpublished), Univ. Pennsylvania, 1984.

177. P. FREYD, D. YETTER, J. HOSTE, W. B. R. LICKORISH, K. C. MILLETT and A. OCNEANU, 'A new polynomial invariant of knots and links', *Bull. American Math. Soc.* **12** (1985) 239–246.

178. P. DE LA HARPE, P. M. KERVAIRE, and C. WEBER, 'On the Jones polynomial', *L'Enseignement Math.* **32** (1986) 271–335.

179. C. F. HO, 'A new polynomial invariant for knots and links — preliminary report', *American Math. Soc. Abstracts* **6** (1985) p. 300.

180. C. F. HO, *On Polynomial Invariants for Knots and Links*, Ph.D. Thesis, California Institute of Technology, 1986.

181. J. HOSTE, 'A polynomial invariant of knots and links', *Pacific J. Math.* **124** (1986) 295–320.

182. V. F. R. JONES, 'A polynomial invariant for knots via von Neumann algebras', *Bull. American Math. Soc.* **12** (1985) 103–111.

183. V. F. R. JONES, 'Hecke algebra representations of braid groups and link polynomials', *Annals of Math.* **126** (1987) 335–388.

184. V. F. R. JONES, *Subfactors and Knots*, CBMS Regional Conference Series **80**, American Math. Soc., 1991.

185. T. KANENOBU, 'Infinitely many knots with the same polynomial invariant', *Proc. American Math. Soc.* **97** (1986) 158–162.

186. T. KANENOBU, 'Examples on polynomial invariants of knots and links', *Math. Annalen* **275** (1986) 555–572.

187. T. KANENOBU, 'Examples on polynomial invariants of knots and links II', *Osaka J. Math.* **26** (1989) 465–482.

188. L. H. KAUFFMAN, 'State models and the Jones polynomial', *Topology* **26** (1987) 395–407.

189. L. H. KAUFFMAN, 'Statistical mechanics and the Jones polynomial', *Braids* (Proc. Conference Santa Cruz 1986), eds. J. S. Birman and A. Libgober, Contemporary Math. **78**, American Math. Soc., 1988, pp. 263–297.

190. L. H. KAUFFMAN, 'An invariant of regular isotopy', *Trans. American Math. Soc.* **318** (1990) 417–471.

191. R. C. Kirby and P. Melvin, 'Evaluations of the 3-manifold invariants of Witten and Reshetikhin–Turaev for $sl(2, \mathbb{C})$', *Inventiones Math.* **105** (1991) 473–545.

192. W. B. R. Lickorish and K. C. Millett, 'The reversing result for the Jones polynomial', *Pacific J. Math.* **124** (1986) 173–176.

193. W. B. R. Lickorish and K. C. Millett, 'A polynomial invariant of oriented links', *Topology* **26** (1987) 107–141.

194. W. B. R. Lickorish and M. B. Thistlethwaite, 'Some links with nontrivial polynomials and their crossing numbers', *Commentarii Math. Helvetici* **63** (1988) 527–539.

195. A. A. Markov, 'Über die freie Äquivalenz der geschlossenen Zöpfe', *Recueil Math. Moscou* **1** (1936) 73–78.

196. H. R. Morton, 'Threading knot diagrams', *Math. Proc. Cambridge Philos. Soc.* **99** (1986) 247–260.

197. H. R. Morton, 'The Jones polynomial for unoriented links', *Quarterly J. Math. Oxford* **37** (1986) 55–60.

198. H. R. Morton and H. B. Short, 'The 2-variable polynomial of cable knots', *Math. Proc. Cambridge Philos. Soc.* **101** (1987) 267–278.

199. H. R. Morton and P. Traczyk, 'The Jones polynomial of satellite links around mutants', *Braids* (Proc. Conference Santa Cruz 1986), eds. J. S. Birman and A. Libgober, Contemporary Math. **78**, American Math. Soc., 1988, pp. 587–592.

200. K. Murasugi, 'Jones polynomials and classical conjectures in knot theory', *Topology* **26** (1987) 395–407.

201. A. Ocneanu, *A Polynomial Invariant for Knots: A Combinatorial and Algebraic Approach*, Preprint (unpublished), MSRI, Univ. California, 1985.

202. J. H. Przytycki and P. Traczyk, 'Invariants of links of Conway type', *Kobe J. Math.* **4** (1987) 115–139.

203. N. Reshetikhin and V. G. Turaev, 'Invariants of 3-manifolds via link polynomials and quantum groups', *Inventiones Math.* **103** (1991) 547–597.

204. R. A. Stong, 'The Jones polynomial of parallels and applications to crossing number', *Pacific J. Math.* **164** (1994) 383–395.

205. M. B. Thistlethwaite, 'A spanning tree expansion of the Jones polynomial', *Topology* **26** (1987) 297–309.

206. M. B. Thistlethwaite, 'Kauffman's polynomial and alternating links', *Topology* **27** (1988) 311–318.

207. M. B. Thistlethwaite, 'On the Kauffman polynomial of an adequate link', *Inventiones Math.* **93** (1988) 285–296.

208. V. G. Turaev, 'A simple proof of the Murasugi and Kauffman theorems on alternating links', *L'Enseignement Math.* **33** (1987) 203–225.

209. V. G. TURAEV, 'The Yang–Baxter equation and invariants of links', *Inventiones Math.* **92** (1988) 527–553.

210. E. WITTEN, 'Quantum field theory and the Jones polynomial', *Communications in Math. Physics* **121** (1989) 351–399.

10   Closed Braids and Arc Presentations

211. J. W. ALEXANDER, 'A lemma on a system of knotted curves', *Proc. National Acad. Sciences USA* **9** (1923) 93–95.

212. Y. BAE and C.-Y. PARK, 'An upper bound of arc index of links', *Math. Proc. Cambridge Philos. Soc.* **129** (2000) 491–500.

213. E. BELTRAMI, *Arc Index of Links*, Ph.D. Thesis, Pisa Univ., 1998.

214. E. BELTRAMI, 'Arc index of non-alternating links', *J. Knot Theory and Its Ramifications* **11** (2002) 431–444.

215. E. BELTRAMI and P. R. CROMWELL, 'Minimal arc-presentations of some non-alternating knots', *Topology and Its Applications* **81** (1997) 137–145.

216. E. BELTRAMI and P. R. CROMWELL, 'A limitation on algorithms for constructing minimal arc-presentations from link diagrams', *J. Knot Theory and Its Ramifications* **7** (1998) 415–423.

217. J. S. BIRMAN and E. FINKELSTEIN, 'Studying surfaces via closed braids', *J. Knot Theory and Its Ramifications* **7** (1998) 267–334.

218. J. S. BIRMAN and W. W. MENASCO, 'Studying links via closed braids IV: composite links and split links', *Inventiones Math.* **102** (1990) 115–139.

219. J. S. BIRMAN and W. W. MENASCO, 'Studying links via closed braids V: closed braid representatives of the unlink', *Trans. American Math. Soc.* **329** (1992) 585–606.

220. J. S. BIRMAN and W. W. MENASCO, 'Special positions for essential tori in link complements', *Topology* **33** (1994) 525–556. Errata: *Topology* **37** (1998) p. 225.

221. H. BRUNN, 'Über verknotete Kurven', *Verhandlungen des Internationalen Math. Kongresses* (Zurich 1897), Leipzig, 1898, pp. 256–259.

222. J. CANTARELLA, X. W. FABER and C. A. MULLIKIN, 'Upper bounds for ropelength as a function of crossing number', *Topology and Its Applications* **135** (2003) 253–264.

223. P. R. CROMWELL, 'Embedding knots and links in an open book I: basic properties', *Topology and Its Applications* **64** (1995) 37–58.

224. P. R. CROMWELL, 'Arc presentations of knots and links', *Knot Theory* (Proc. Conference Warsaw 1995), eds. V. F. R. Jones *et al.*, Banach Center Publications **42**, 1998, pp. 57–64.

225. P. R. CROMWELL and I. J. NUTT, 'Embedding knots and links in an open book II: bounds on arc index', *Math. Proc. Cambridge Philos. Soc.* **119** (1996) 309–319.

226. I. A. Dynnikov, *Arc Presentations of Links: Monotonic Simplification*, Preprint, Moscow State Univ., 2002. e-Print archive: math.GT/0208153.

227. T. Erlandsson, *Geometry of Contact Transformations in Dimension Three*, Ph.D. Thesis, Uppsala Univ., 1981.

228. J. B. Etnyre and K. Honda, 'Knots and contact geometry I: torus knots and the figure eight knot', *J. Symplectic Geometry* **1** (2001) 63–120.

229. J. Franks and R. F. Williams, 'Braids and the Jones–Conway polynomial', *Trans. American Math. Soc.* **303** (1987) 97–108.

230. H. C. Lyon, 'Torus knots in the complements of links and surfaces', *Michigan Math. J.* **27** (1980) 39–46.

231. H. R. Morton, 'An irreducible 4-string braid with unknotted closure', *Math. Proc. Cambridge Philos. Soc.* **93** (1983) 259–261.

232. H. R. Morton, 'Seifert circles and knot polynomials', *Math. Proc. Cambridge Philos. Soc.* **99** (1986) 107–110.

233. H. R. Morton and E. Beltrami, 'Arc index and the Kauffman polynomial', *Math. Proc. Cambridge Philos. Soc.* **123** (1998) 41–48.

234. K. Murasugi, 'On the braid index of alternating links', *Trans. American Math. Soc.* **326** (1991) 237–260.

235. L. P. Neuwirth, '*-Projections of knots', *Algebraic and Differential Topology: Global Differential Geometry*, ed. G. M. Rassias, Teubner-Texte zur Math. **70**, 1984, pp. 198–205.

236. K. Y. Ng, 'Essential tori in link complements', *J. Knot Theory and Its Ramifications* **7** (1998) 205–216.

237. I. J. Nutt, *Braid Index of Satellite Links*, Ph.D. Thesis, Liverpool Univ., 1995.

238. I. J. Nutt, 'Arc index and the Kauffman polynomial', *J. Knot Theory and Its Ramifications* **6** (1997) 61–77.

239. I. J. Nutt, 'On the braid index of satellite links', *Math. Proc. Cambridge Philos. Soc.* **126** (1999) 77–98.

240. C. Park and M. Seo, 'On the arc index of an adequate link', *Bull. Australian Math. Soc.* **61** (2000) 177–187.

241. L. Rudolph, 'Quasipositive annuli (constructions of quasipositive knots and links IV)', *J. Knot Theory and Its Ramifications* **4** (1992) 451–466.

242. P. Traczyk, 'A new proof of Markov's braid theorem', *Knot Theory* (Proc. Conference Warsaw 1995), eds. V. F. R. Jones *et al.*, Banach Center Publications **42**, 1998, pp. 409–419.

243. P. Vogel, 'Representation of links by braids: a new algorithm', *Commentarii Math. Helvetici* **65** (1990) 104–113.

244. S. Yamada, 'The minimal number of Seifert circles equals the braid index of a link', *Inventiones Math.* **89** (1987) 347–356.

Appendices

245. J. A. CALVO, *Geometric Knot Theory*, Ph.D. Thesis, Univ. California Santa Barbara, 1998.

246. T. KANENOBU and H. MURAKAMI, 'Two-bridge knots with unknotting number one', *Proc. American Math. Soc.* **98** (1986) 499–502.

247. P. KRONHEIMER and T. MROWKA, 'Gauge theory for embedded surfaces I', *Topology* **32** (1993) 773–826.

248. P. KRONHEIMER and T. MROWKA, 'Gauge theory for embedded surfaces II', *Topology* **34** (1995) 37–97.

249. W. B. R. LICKORISH, 'The unknotting number of a classical knot', *Combinatorial Methods in Topology and Algebraic Geometry*, Contemporary Math. **44**, American Math. Soc., 1985, pp. 117–121.

250. C. L. MCCABE, 'An upper bound on the edge numbers of 2-bridge knots and links', *J. Knot Theory and Its Ramifications* **7** (1998) 797–805.

251. Y. NAKANISHI, 'A note on unknotting number', *Math. Seminar Notes Kobe Univ.* **9** (1981) 99–108.

252. P. OZSVÁTH and Z. SZABÓ, *Knots with Unknotting Number One and Heegaard Floer Homology*, Preprint, Institute for Advanced Study, Princeton, 2004. e-Print archive: `math.GT/0401426`.

253. A. STOIMENOW, *Polynomial Values, the Linking Form and Unknotting Numbers*, Preprint, Max Planck Institut, 2000. e-Print archive: `math.GT/0405076`.

Internet Resources

Many recent papers listed above can be found in the Geometric Topology section of the e-Print archive: `http://www.arxiv.org`.

254. J. HOSTE and M. B. THISTLETHWAITE, *Knotscape* (a graphical interface to the knot database), `http://www.math.utk.edu/~morwen`.

255. H. R. MORTON and H. B. SHORT, Pascal source code for computing Homfly polynomials from braids, `http://www.liv.ac.uk/~su14`.

Index

When page numbers are in italic a figure has been indexed.